CHARGE TRANSFER
INTERACTIONS
OF BIOMOLECULES

ב״ה

CHARGE TRANSFER
INTERACTIONS
OF BIOMOLECULES

Michael A. Slifkin

Department of Pure and Applied Physics
University of Salford, Salford, Lancs, England

1971

ACADEMIC PRESS · LONDON · NEW YORK

CHEMISTRY

ACADEMIC PRESS INC. (LONDON) LTD.
24-28 Oval Road,
London, NW1 7DX

U. S. Edition published by
ACADEMIC PRESS INC.
111, Fifth Avenue,
New York, New York 10003.

Copyright © 1971 By ACADEMIC PRESS INC. (LONDON) LTD.

Library of Congress Catalog Card Number: 77-170744
International Standard Book Number: 0-12-648850-9

PRINTED IN GREAT BRITAIN BY
THE WHITEFRIARS PRESS LTD., LONDON AND TONBRIDGE

Preface

This book has a dual purpose. I have attempted to review all the relevant work on the topic and to present it in a coherent manner. Wherever possible numerical results have been assembled in tables so that comparisons can be made between different compounds. In the chapter on experimental techniques, I have tried to give sufficient summary of the various techniques available to enable anyone contemplating work in this field to decide which technique would best suit his problem. Inevitably those techniques with which I am myself familiar have been treated in much greater detail, but here I can pass on something of my own experience.

The subject matter of this book is surrounded by much controversy and I have attempted to be fair to authors holding different views from my own although I have not hesitated in putting forward my own views.

Modern opinion takes the view that hydrogen bonding is a special case of charge transfer complexing, an opinion with which I concur, but, as this is a well recognized phenomenon in its own right, it has not been specifically included, but only where it is relevant to the main topic. Interactions involving metallic ions, although sometimes classed as charge transfer interactions, have been excluded, and also chelates, as these are different kinds of interaction.

Much work has been carried out on charge and electron transfer processes in biological systems, tissue extracts, mitochondria, etc. These are such complicated systems that as a physicist, I do not feel competent to review this field. Instead the discussion will be restricted to fairly simple *in vitro* systems. The relevance of the work discussed, to actual biological systems will not be commented on in depth as this requires deeper knowledge of biology.

Chapter 1 is a general discussion of charge transfer and charge transfer complexes, Chapter 2 deals with experimental methods and techniques, and the remaining chapters deal with charge transfer interactions of different families of related molecules. These is inevitably some cross referencing required as many of the compounds listed under separate headings interact with each other.

v

I have attempted to select material comprehensively but it is more than likely that some work has been overlooked especially in the enormous flavin field. For this I must offer my apologies to the reader for not giving a complete account and to my fellow workers for inadvertently overlooking their contribution to the subject.

M. A. Slifkin   *August 1971*

Acknowledgements

I should like to express my gratitude to the many authors who have allowed me to reproduce diagrams and tables from their published work and to the following who have granted me permission to reproduce copyright items from their books and journals:

American Chemical Society, Springer-Verlag, Pergamon Press Ltd., National Academy of Sciences, University of Tokyo Press, Elsevier Publishing Co., American Institute of Physics, American Pharmaceutical Association, The American Society of Biological Chemists Inc., Berichte der Bunsen-Gesellschaft fur Physikalische Chemie, The Biochemical Journal, The Chemical Society, Commonwealth Science and Industrial Research Organisation, Experientia, The Faraday Society, Helvetica Chimica Acta, The Japanese Biochemical Society, H. K. Lewis and Co. Ltd., Longman Group Ltd., Macmillan (Journals) Ltd., North-Holland Publishing Co., The Royal Institute of Chemistry, Academic Press Inc., Indian Journal of Chemistry, John Wiley and Sons Inc., The Japanese Pharmaceutical Society.

M. A. S.

Contents

Chapter 3: Amino Acids and Proteins

Chapter 4: Purines and Pyrimidines

Chapter 5: Indoles and Imidazoles

Chapter 6: Porphyrins

List of Tables

Introduction: The Role of
Charge Transfer Complexes in Biology

There is no clear evidence of any role played by charge transfer complexes in biology. Charge transfer complexes do however possess certain properties which could be important in biological systems.

The first most obvious property is the transfer of charge from one molecule to another. Electron and charge transfer or transport systems are vitally important in biology. The mechanisms of photosynthesis and oxidative phosphorylation are the object of much intense study. The biochemistry of these systems is still very unclear but charge transfer complexing is a possible method by which molecules in such processes could receive and pass on charge. There is no *in vivo* evidence for such complexes although many molecules or their analogues, thought to be active in those systems, can form charge transfer complexes *in vitro*.

Another property of charge transfer complex forces is that they are relatively long range as compared to chemical forces. Thus, typical distances between molecules in these complexes are 3.2 to 3.4 Å whereas chemical bond lengths are less than about 1.5 Å. Charge transfer would therefore facilitate interaction between mobile molecules over so comparatively long separations.

Charge transfer forces usually hold the components in the complex together in specific orientation. This could perhaps act to bring large biomolecules together with their prosthetic groups correctly aligned for interaction

Charge transfer complexes normally have enthalpies of dissociation of less than 8 kcal/mole. This is the energy of the energy rich phosphate bonds in ATP and ADP. This might be quite fortuitous as there is no evidence that ATP can interact with charge transfer complexes losing a phosphate and supplying the energy to split the complex. It is an interesting speculation whether this figure of 8 kcal/mole might have some significance in biological charge transfer complexes.

Many large biomolecules are good semiconductors. One could speculate that electron transfer through highly organised systems such as the mitochondria is perhaps by conduction. The role of charge transfer complexes could be to control and modify the electron flow like the base of a transistor. Charge transfer complexes in general have much lower activation energies and much higher conductivities than the free components. The effect of introducing a charge donor or charge acceptor into an electron conducting system of biomolecules would be very similar to introducing electron donating or hole donating impurities into inorganic semiconductors.

One other property of charge transfer complexes which is only likely to be of importance on the surface of biosystems is their colour. Sometimes charge transfer complexes possess absorption bands to the long wave length side of the light absorption of the free components. Hence complexing allows the absorption of light energies which neither of the components could absorb on their own.

CHAPTER 1

Charge Transfer and Charge Transfer Complexes

1.1 Introduction

All molecules are built up from atoms. The molecule can be considered as held together by electrons from the atoms circling all or part of the molecule. In a molecule, especially the large molecules that are normally encountered in biological systems, one finds regions of the molecules which are electropositive and other regions which are electronegative; or one may have molecules which relative to others are electropositive or electronegative. These inequalities in electropotential can lead to electron sharing or electron donation and acceptation between molecules or even parts of molecules. Although one can thus talk about electron donors and electron acceptors, these are relative terms. Whereas polycyclic aromatic hydrocarbons will donate charge to benzoquinones and are thus charge donors, they accept charge from sodium in which case they are charge acceptors. Conversely carbonyl groups are acceptors from the same hydrocarbons but are donors to bromine. There is no reason why any molecule should not be a charge donor and acceptor under the appropriate conditions. Nevertheless, this is not always so. Before a charge interaction can take place, the molecules or parts of molecules must be in sufficient proximity to each other so that the difference in electropotential can be recognized. With many molecules, particularly large molecules, steric hindrance may well prevent such close approach. Other kinds of physical forces may be operating between the molecules, preventing their approach. In many cases, molecules may chemically interact so that even though conditions are very favourable for a charge sharing or transfer to take place, the effects are overtaken by the chemical changes. Often the conditions which are favourable for charge transfer, are the very conditions which favour chemical changes.

Molecules do not exist in isolation, except in outer space, and no inter-action between molecules can be divorced from the effects of the immediate surroundings. Polar environments will facilitate charge transfer which results in an increase in dipole moment of the two molecules donating and accepting as compared to non-polar environments. In media of low viscosity, Brownian motion which causes the molecules to move about will be a hindrance to charge transfer between molecules. As the temperature decreases, Brownian motion decreases and there will be less inhibition of the charge transfer. In a rigid matrix, often a frozen solution, the molecules may be held in such close proximity that charge transfer will be greatly facilitated. In rigid matrices, the molecules will not experience any Brownian motion, so that charge transfer interactions will always be greatly enhanced. The optimum distance for charge transfer between sites is of the order of $3 0 \sim 3.4$ Å. Indeed molecules which seem not to interact in solution may do so when ground together in a solid state. A good example of this is quinhydrone which is actually an equimolar mixture of hydroquinone and benzoquinone. Although benzoquinone is dark and hydroquinone is white, grinding the two together produces a purple substance. However on dissolving, this substance gives colourless solutions unless frozen down to low temperatures when the purple colour appears but then disappears on warming again.

There is a special case of charge transfer interaction of which quinhydrone is a good example which causes the formation of weak molecular complexes, which are believed by some to have special significance in biology. These complexes are called charge transfer complexes. This term has been widely applied to many systems and is the subject of much controversy.

1.2. Charge Transfer Complexes

The term charge transfer complex was first coined by Mulliken (1) to describe a certain type of complex with distinctive features. It had been known for a very long time that mixtures of certain molecules or classes of molecules, a good example being quinhydrone, could form highly coloured mixtures in the solid or in solution, although the coloured mixture seemed not to be a new chemical entity but possessed the chemical properties of the components. Other changes associated with these mixtures were a marked solubilization of the components, a decrease in diamagnetic susceptibility and increase in paramagnetic susceptibility of the mixtures as compared to the sum of these parameters of the components. Often the infra-red spectrum of the mixture was little different from the sum of the individual components. One striking property of these mixtures was the ability to form crystals having regular simple stoichiometry and structure (2). Although there had been several theoretical attempts made to explain these phenomena, undoubtedly the most successful is the one put forward by Mulliken (1).

1.3. Mulliken Theory of Charge Transfer Complexes

According to Mulliken, the phenomena just described arises from the formation of weak complexes, enthalpies of dissociation being typically of the order of a few kcal/mole, between two molecules, one an electron acceptor and the other an electron acceptor relative to each other. The complex exists in two states, a ground state and an excited state. In the ground state, the two molecules experience the normal physical forces one would expect from two molecules in close proximity i.e. van der Waals forces etc. and in addition a small amount of charge is transfered from the donor to the acceptor which contributes some additional binding energy to the complex. The excited state is promoted when the ground state complex absorbs light of suitable energy. In this excited state the electron which was only slightly shifted towards the acceptor is almost wholly transfered. It is the transfer of the electron on the absorption of light which gives the characteristic colours of these complexes.

Mulliken has explained this using a valence-bond model. The ground state of the complex has a wavefunction ψ_N which is a hybrid of two wave-functions $\psi_{(A, D)}$ and $\psi_{(A-D+)}$. $\psi_{(A, D)}$ is the no-bond function and is the wavefunction of the two molecules in close proximity with no charge transfer between them. However it can include contributions from classical electro-static forces, van der Waals forces and various dispersion forces and dipole interactions. $\psi_{(A-D+)}$ is called the dative function and is the wave-function of the two molecules bound together by an electron being totally transfered from the donor D to the acceptor A.

The ground state of the complex is described thus:

$$\psi_N = a\psi_{(A, D)} + b\psi_{(A-D+)} \quad \text{where } a \gg b.$$

The excited state of the complex is described thus:

$$\psi_E = b^*\psi_{(A-D+)} - a^*\psi_{(A, D)} \quad \text{where } b^* \gg a^*.$$

The energy levels of the complex can be found by solving the Schroedinger equation

$H\psi_N = W\psi_N$, where H is the Hamiltonian operator and W is the energy

or

$$(c\psi_{1(A, D)} + c\psi_{2(A-D+)})H - W = 0,$$

using a generalized wavefunction. This is solved by the usual methods using the Ritz variational method to yield the secular equations

$$c_{11}(H_{00} - W) + c_{21}(H_{01} - SW) = 0$$

and
$$c_{12}(H_{10} - SW) + c_{22}(H_{11} - W) = 0$$

which is only non-trivial if the secular determinant

$$\begin{vmatrix} H_{00} - W, & H_{01} - SW \\ H_{10} - SW, & H_{11} - W \end{vmatrix} = 0$$

where
$$H_{00} = \int \psi_{(A,\,D)} H \psi^\dagger_{(A,\,D)}\, d\tau$$
$$H_{01} = \int \psi_{(A,\,D)} H \psi_{(A^-D^+)}\, d\tau$$
$$H_{10} = \int \psi_{(A^-D^+)} H \psi_{(A,\,D)}\, d\tau$$
$$H_{11} = \int \psi_{(A^-D^+)} H \psi_{(A^-D^+)}\, d\tau$$

and
$$S = \int \psi_{(A,\,D)} \psi_{(A^-D^+)}\, d\tau \text{ (the overlap-integral).}$$

The solution of this equation gives two roots for the energy W, the lower one corresponding to the ground state and the upper one to the excited state of the complex.

These roots are

$$W = W_N = W_0 - \frac{(H_{01} - W_0 S)^2}{W_1 - W_0} \text{ for } S^2 \ll 1$$

and
$$W = W_E = W_1 + \frac{(H_{01} - W_1 S)^2}{W_1 - W_0}$$

where we have replaced H_{00} by W_0 and H_{11} by W_1. These terms represent the energies of the D, A and $D^+ A^-$ states respectively. The difference between the two states of the complex gives the energy of the charge transfer transition $h\nu_{CT}$, thus

$$h\nu_{CT} = W_E - W_N = W_1 - W_0 + \frac{(H_{01} - W_1 S)^2 + (H_{01} - W_0 S)^2}{W_1 - W_0}.$$

The expression $W_1 - W_0$ is given by $I_D - E_A - \Delta$, where I_D is the ionization potential of the donor, i.e. the energy to remove the least bound electron to infinity, E_A is the electron affinity of the acceptor, i.e. the energy gained when an electron is taken from infinity and placed in the lowest unoccupied orbital of the acceptor and Δ represents other terms, the major one being the Coulombic interaction energy between the now donor and acceptor ions and is of the form e/r_{AD}. In solution there will also be solvation terms but provided one works with the same solution these can usually be ignored for families of related molecules.

Therefore

$$h\nu_{CT} = I_D - E_A - \Delta + \frac{(H_{01} - W_1 S)^2 + (H_{01} - W_0 S)^2}{I_D - E_A - \Delta}.$$

The numerator in the right hand term is approximately constant and usually written as $2\beta^2$. Hence there is a parabolic relationship between the energy of the charge transfer band and the ionization potential for complexes between different donors and a common acceptor. For weakly bound

† Indicates the complex conjugate. This is not strictly required here as the wavefunctions are real.

complexes, H_{01} and S hence $2\beta^2$ is small giving the linear relationship

$$hv_{CT} = I_D - E_A - \Delta.$$

Thus for a series of weak complexes of related donors with a common acceptor the ionization potentials should linearly correlate with the positions of the charge transfer bands.

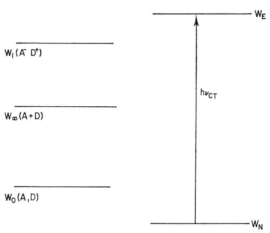

Fig. 1.1. Energy level diagram for a charge transfer complex.

W_∞ = energy of the two molecules isolated from each other.
W_1 = energy of molecules in dative mode.
W_0 = energy of molecules in no-bond mode.
W_E = energy of excited state of complex.
W_N = energy of ground state of complex.
$W_\infty - W_N$ = total binding energy of charge transfer complex.
$W_0 - W_N$ = binding energy due to charge transfer interaction.
$W_\infty - W_0$ = binding energy due to van der Waals and dispersion forces etc.
    (energies are not drawn to scale).

From Fig. 1.1 it is seen that the charge transfer binding energy is given by

$$W_0 - W_N = \frac{(H_{01} - W_0 S)^2}{W_1 - W_0} \text{ which is approximately } - \left(\frac{b}{a}\right)^2 W_1 - W_0.$$

The binding energy should therefore correlate with $I_D - E_A - \Delta$ provided other interaction terms are small.

The ionization potentials of many molecules have not been experimentally determined. However, estimates of their ionization potentials have frequently been carried out using molecular orbital theory (3). Ionization potentials as such have not been estimated, rather the energy of the highest filled molecular orbital is given. Most of the molecules of biological interest possess conjugated π-electron systems which lend themselves readily to calculation. What is calculated by fairly gross methods of approximation are the energies of the molecular orbitals in terms of parameters α and β (not the β of the previous paragraph) and a fraction x_i which can be positive or negative. The energy E_i is given by $E_i = \alpha + x_i\beta$. These parameters α and β are either determined empirically or left undetermined. Having established the energies of the orbitals, the electrons are imagined placed in them, two to each orbital as required by the Pauli Exclusion Principle. The theory predicts

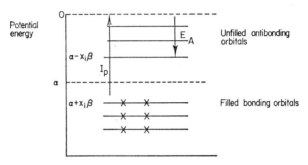

Fig. 1.2. Energy level diagram of typical conjugated ring system.

more orbitals than half the number of electrons. Usually the filled orbitals are bonding orbitals, that is, filling of these orbitals with electrons adds stability to the molecule. The unfilled orbitals of the ground state are usually anti-bonding orbitals so that filling of these orbitals by excited or donated electrons lowers the stability of the molecule. The ionization potential is the energy required to remove the least bound molecule to infinity and in terms of the energy diagram is equal to $\alpha + x_i\beta$ for the highest filled orbital. The electron affinity is the gain in energy on bringing an electron from infinity to the molecule which must then occupy the lowest empty orbital. The electron affinity is given by $\alpha - x_i\beta$ for the lowest occupied orbital. Because of the uncertainties in the values of α and β, ionization potentials and electron affinities are often expressed in terms of the coefficients x_i.

Substitution of the values calculated for W in the secular equations give the following ratios of the coefficients of the wavefunctions:

$$\frac{b}{a} = -\frac{(H_{01} - W_0 S)}{W_1 - W_0} \text{ for } \psi_N \quad \text{and} \quad \frac{b^*}{a^*} = -\frac{(H_{01} - W_1 S)}{W_1 - W_0} \text{ for } \psi_E$$

where $c_{11} = a$ $c_{21} = b$ $c_{12} = -b^*$ $c_{22} = a^*$.

These parameters can be obtained from the dipole moments of the molecules and their complex. The dipole moment of the complex in the ground state is given by

$$\mu_N = \int \psi_N \mu \psi_N \, d\tau = a^2 \mu_0 + b^2 \mu_1 + 2ab\mu_{01} \quad \text{where } \mu = -e\sum r$$

$$\mu_0 = \int \psi_{(A, D)} \mu \psi_{(A, D)} \, d\tau,$$

$$\mu_{01} = \int \psi_{(A, D)} \mu \psi_{(A-D^+)} \, d\tau$$

and $$\mu_1 = \int \psi_{(A-D^+)} \mu \psi_{(A-D^+)} \, d\tau$$

μ_0 is the vector sum of the individual dipole moments of molecules D and A. If, as often will be the case in the complexes under consideration, μ_0 is zero then

$$\mu_N = \mu_1(b^2 + abS) \quad \text{as } \mu_{01} \approx \tfrac{1}{2}\mu_1 S.\dagger$$

The normalization equations $\int \psi_N \psi_N \, d\tau = \int \psi_E \psi_E \, d\tau = 1$ and the orthonomality condition that $\int \psi_N \psi_E \, d\tau = 0$ gives

$$a^2 + 2abS + b^2 = a^{*2} - 2a^*b^*S + b^{*2} = 1$$

and $a^*(b + aS) = b^*(bS + a)$.

Thus estimation or measurement of S and the dipole moments μ_N and $\mu_1\ddagger$ will enable the coefficients to be calculated.

A relationship can be derived between these coefficients and the extinction coefficient ε_{CT} of the charge transfer band. The extinction coefficient is defined from the Beer-Lambert law governing the absorption of light in a medium thus

$$I = I_0 e^{-\varepsilon_{CT}cl},$$

where I_0 is the intensity of light before entering the medium, I is the emerging intensity, c is the concentration in moles and l is the absorption length in cm.

The extinction coefficient is related to the theoretical oscillator strength f by

$$f = 4.32 \times 10^{-9} \int \varepsilon_{CT} \, dv.$$

$\int \varepsilon_{CT} \, dv$ is the area under the curve of the extinction coefficient of the absorption band in question $vs.$ the frequency. To a first approximation

$$f = 4.32 \times 10^{-9} \varepsilon_{max} \Delta v_{\frac{1}{2}}$$

where ε_{max} is the maximum extinction coefficient of the band and $\Delta v_{\frac{1}{2}}$ is the half-width, i.e. the width of the band at half the maximum extinction.

The extinction coefficient is related to the transition dipole by

$$\mu_{EN} = .0958 \left[\frac{\varepsilon_{max} \Delta v_{\frac{1}{2}}}{\tilde{v}} \right]^{\frac{1}{2}}$$

where $\tilde{v} \approx v$ at ε_{max} and μ_{EN} is defined as $-e \int \psi_E \sum_i r_i \psi_N \, d\tau$.

μ_{EN} is also given by

$$(a^*b\mu_1 - ab^*\mu_0) + (aa^* - bb^*)\mu_{01} = a^*b(\mu_1 - \mu_0) + (aa^* - bb^*)(\mu_{01} - S\mu_0),$$

and for the case of $\mu_0 = 0$

$$\mu_{EN} \approx (a^*b)\mu_1 + (aa^* - bb^*)\mu_{01}$$

† The dipole moment μ_{01} is equal to the moment generated when an amount of charge Se is transferred from the donor to somewhere between the donor and acceptor. Hence $\mu_{01} \simeq \tfrac{1}{2}\mu_1 S$.

‡ μ_1 is approximately er_{AD}/E where r_{AD} is the separation between D and A in the complex and E is the dielectric constant.

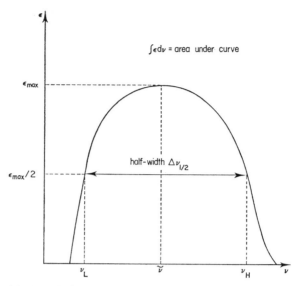

Fig. 1.3. Hypothetical absorption band illustrating various parameters.

but as before

$$\mu_{01} \approx \tfrac{1}{2}\mu_1 S \quad \text{then} \quad \mu_{EN} = [a^*b + \tfrac{1}{2}S(aa^* - bb^*)]\mu_1.$$

Hence the extinction coefficient should increase with increased strength of complexing.

This relationship has been shown to hold in some cases but not in other (2). Departure from this relationship is probably due to the inadequacies of the methods used to obtain the extinction coefficients of the charge transfer transitions (Section 1.6).

For very weak complexes, although b may be very small, the right-hand term may still be appreciable thus explaining the often very strong absorption.

An alternative explanation is due to Murrell (4) who has suggested that intensity of these charge transfer transitions can be increased by borrowing intensity from the strong intrinsic transitions of the donor or acceptor. The excited state wavefunction ψ_E is then given by

$$\psi_E = a^*\psi_{(A^- D^+)} - b^*\psi_{(A, D)} - c^*\psi_{(A, D^*)} - d^*\psi_{(A^*, D)}$$

Such intensity borrowing could only occur in the case of contact charge transfer, i.e. where there is a momentary exchange of charge when the molecules approach but no binding. The reason that intensity borrowing cannot occur with bound complexes is that the charge transfer transition is polarized along the axis joining the donor to the acceptor whereas for aromatic molecules the $\pi-\pi$ transitions are polarized perpendicularly to this axis and there is therefore no overlap and terms of the form

$$\int \psi_{(A^-D^+)} \psi_{(A^*,D)} \, d\tau$$

etc. do not exist.

1.4. Nomenclature

The nomenclature has been subject to a certain amount of criticism (5). As is obvious from the description, the major binding forces in the ground state are not necessarily charge transfer forces. Indeed the so-called no-bond function contains terms describing dispersion forces etc. which may constitute almost the whole of the complexing forces. It has been shown for many complexes, especially those between aromatic hydrocarbons and different acceptors, that very little binding can be attributed to charge transfer forces (5). The objection is therefore raised to the terminology that although these are complexes displaying charge transfer spectra, the binding arises from a different mechanism. Therefore, the term charge transfer complex is misleading in the case of those complexes in which definite evidence of charge transfer stabilization in the ground state is lacking. On the other hand other authors restrict the term charge transfer complex to complexes displaying charge transfer bands (6) irrespective of the ground state binding forces. They reject its use for many of the complexes described herein which are thought to involve charge transfer in the ground state although charge transfer bands as such have not been found.

The term electron-donor-acceptor complex has been used frequently as synonymous with charge transfer complex (2, 7). Some authors object to this in cases other than where there is definite evidence of strong charge transfer in the ground state (5). Bent (8) in a recent review has used this term to cover all the types of complexes discussed above. One criterion given in Bent's review for all electron donor acceptor complexes is that the intermolecular separation of the partners in the complex is less than one would expect if only dispersion forces were present.

Wallwork (9) has measured the intermolecular distance between the components of several charge transfer complexes of the weak type similar to those thought by Dewar and Thompson (5) not to have any charge transfer in the ground state. Wallwork has found that the intermolecular distances are between 0.1 and 0.2 Å shorter than the normal separation of molecules held together in crystals by van der Waals forces.

A more fundamental objection has been raised to the whole concept of charge transfer complexes as put forward by Mulliken by Dewar who states that all the various weak interaction forces such as dipole–dipole interaction, mutual polarization, etc., can be described in terms of resonance structures which could all include dative functions of the kind used by Mulliken and that there is no difference between these different weak interactions (5). Dewar therefore proposes that no attempt need be made to differentiate between different complexing forces and the the weak complexes may all be called complexes without further qualification. This view is contested by Mulliken and Person who whilst accepting that charge transfer stabilization may be small in the case of complexes of aromatic hydrocarbons and conventional acceptors such as chloranil and tetracyanoethylene state that it is by no means wholly insignificant and that furthermore the other types of weak interactions can be described in terms of hybrid structures which do not include the dative function whereas those complexes termed by them charge transfer complexes cannot (7).

Foster in the most recent book in this field points out the difficulties of nomenclature and uses the term charge transfer to cover all weak complexes because of its wide use in the literature (2).

1.5. Classification of Complexes

Mulliken has listed a large number of different types of charge transfer complexes (1). For our purposes only three broad classes of donors need be considered.

The first class is where the donor is a π-electron from a conjugated system; polycyclic aromatic hydrocarbons are typical members of this group. These molecules are usually weak electron donors. Changes in the infra-red spectra of complexes of these donors are generally very small (10). It is this class to which doubt has been expressed as to whether there is any charge transfer in the ground state (5).

The second class of donor is that in which the donated electron is an n-electron or lone pair electron most frequently located on the nitrogen in an amino group, although donation from lone-pair electron on sulphur and oxygen are known. n-Donation can take place from the nitrogen in a heterocyclic ring. These complexes tend to be stronger than the first class. The infrared spectra of these donors in complexes often show marked changes. The carbonyl band of chloranil when complexed with the n-donor glycine shifts to the red by 60 cm^{-1} (1, 12). Some of these complexes are completely dative in the ground state and exhibit the characteristic spectra of the ions.

The third class is less easy to define. There appears to be some evidence that charge transfer can take place between localized regions of positive charge and negative charge in adjacent molecules (Section 1.3 and see Sections 5.9 and 7.9.6). Presumably the difference between this type of interaction and classical electrostatic attraction between charged species is that there is some movement of the charge from one region to another, apart from any movements in the molecular skeletons. This is usually called charge complimentarity.

Acceptors may also be classified and for our purposes we may recognize π-acceptors in which the donated electron goes into a π-orbital of the acceptor, σ-acceptors and localized acceptors analogous to the localized donors.

1.6. Parameters of Charge Transfer Complexes

The most important parameter of these complexes, is the strength or stability of the complex. The most usual parameter quoted in the literature is the association constant, often called the stability constant or equilibrium constant. The inverse of this, the dissociation constant, is found in the earlier literature.

The association constant is defined as follows:

assuming the interaction $\quad D + A \underset{k_2}{\overset{k_1}{\rightleftharpoons}} (D:A)$

where $(D:A)$ is the complex, then

$$K = \frac{[(D:A)]}{[D_0 - (D:A)][A_0 - (D:A)]}$$

where $[D_0]$ and $[A_0]$ are the initial concentrations before interaction. The square brackets denote concentrations. It is usual to express the concentration in molarities in which case K has the units of litres per mole (M^{-1}) and is then designated K_c. If concentrations are expressed in mole fractions then the association constant is designated K_x.

The above relationship presumes that the solvent plays no part in the interaction, a perhaps dubious assumption discussed by Carter, Murrell and Rosch (14). It also assumes that the activity coefficients of the species are unity, an assumption discussed by Rosenthal who would prefer the use of the term equilibrium quotient instead of the association constant, if the activation coefficients are not evaluated (15).

1.7. The Benesi-Hildebrand Equation

The most popular method of evaluating K_c for the interaction is that due to Benesi and Hildebrand (16) who derived the following relationship, for two molecules complexing according to the foregoing scheme, the extinction coefficient of the charge transfer band being given by

$$\varepsilon_{CT} = \frac{\log (I_0/I)}{l[(D:A)]}$$

provided that neither A nor D absorb in the region of the charge transfer transition. l is the pathlength of complex, log (I_0/I) is the optical density of the complex and square brackets represent concentrations. Elimination between this and the previous equation gives,

$$\frac{[A]l}{\log (I_0/I)} = \frac{1}{K\varepsilon_{CT}[D-(D:A)]} + \frac{1}{\varepsilon_{CT}}$$

for $[D] \gg [A]$ this becomes,

$$\frac{[A]l}{\log (I_0/I)} = \left(\frac{1}{K\varepsilon_{CT}}\right)\left(\frac{1}{[D]}\right) + \left(\frac{1}{\varepsilon_{CT}}\right).$$

A plot of $[A]l/[\log (I_0/I)]$ $vs.$ $1/[D]$ for constant $[A]$ and varying $[D]$ will have a slope of $1/K\varepsilon_{CT}$ and an intercept of $1/\varepsilon_{CT}$.

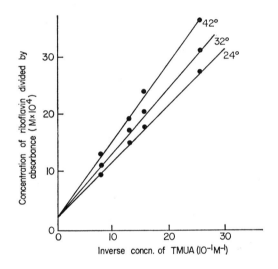

Fig. 1.4. Typical Benesi-Hildebrand plots for the tetramethyluric acid riboflavin complex in water. Absorbances measured at 20.4 kK. Reproduced with permission from Fig. 3, Slifkin, M. A. (1965). *Biochim. Biophys. Acta*, **109**, 617.

The Benesi-Hildebrand equation can be rearranged as to whether one wishes to extrapolate through regions of increasing or decreasing donor concentration. A rearranged equation by Scott (17) is

$$\frac{l[A][D]}{\log (I_0/I)} = \frac{1}{\varepsilon_{CT}}[D] + \frac{1}{K\varepsilon_{CT}}.$$

A plot of the left hand term $vs.$ $[D]$ has a slope of $1/\varepsilon_{CT}$ and an intercept of $1/K\varepsilon_{CT}$.

The practical difference between these two equations is that in the Benesi-Hildebrand equation one extrapolates from a region of decreasing concentration and hence further from the condition that $[D] \gg [A]$, whereas with the Benesi-Hildebrand-Scott equation one extrapolates through a region of decreasing concentration, i.e. decreasing absorbance so that the percentage errors are greater. Whilst the values of $K\varepsilon_{CT}$ are probably correct, the errors in K and ε_{CT} can be quite large (2).

The association constants of complexes which are solvated can be obtained using the equation derived by Carter, Murrell and Rosch (14). For the equilibrium

$$AS_n + DS_m \rightleftharpoons (D:A:S_p) + qS$$

where

$$q = n + m - p$$

and S is the solvent, the equilibrium constant is given by

$$K = \frac{[C]x_s^q}{[AS_n][DS_m]}$$

where x_s is the mole fraction of the free solvent and $[C]$ the concentration of the complex. This equilibrium constant K is related to the one from the Benesi-Hildebrand equation K_{B-H} by

$$K_{B-H} = K - q(m+1)/S_0$$

where S_0 is the total concentration of solvent in the absence of donor.

$$\varepsilon_{B-H} = \varepsilon K/K_{B-H}$$

hence $K\varepsilon_{CT}$ will be correct even though ε_{CT} and K separately might be grossly in error.

Mulliken and Person (2) have given an alternative treatment. Assuming that the true thermodynamic equilibrium constant K_a is related to the measured one by

$$K_a = K_c\Gamma_c,$$

the difference arising because the presence of a large amount of donor effectively changes the nature of the solvent and further assuming a relationship between them

$$\Gamma_c = k_0(1 + c[D]),$$

then the Benesi-Hildebrand equation can be rewritten as

$$\frac{[A]l}{\log(I_0/I)} = \frac{k_0}{K_a\varepsilon_{CT}[D]} + \left(\frac{k_0}{K_a}c + 1\right)\frac{1}{\varepsilon_{CT}}.$$

The difference between the true extinction coefficient ε and the one derived by the Benesi-Hildebrand equation ε_{B-H} is

$$\varepsilon_{B-H} = \varepsilon\left(\frac{1}{[c/K_a]+1}\right) = \frac{K_a}{K_{B-H}}\varepsilon.$$

The difference between the true equilibrium coefficient K_a and the one derived using the Benesi-Hildebrand equation is

$$K_{B-H} = K_a + c.$$

This treatment also shows that $K_{B-H}\varepsilon_{B-H}$ can be correct whilst the individual values be seriously in error.

One variation of the Benesi-Hildebrand equation is of particular interest as it applies in a situation where there are three interacting species (18). Very frequently, biomolecules are studied in buffer solutions which contain anion and cations which can also act as charge donors and acceptors.

For the double simultaneous interactions:

$$A + B \overset{K}{\rightleftharpoons} (A:B), \qquad A + C \overset{G}{\rightleftharpoons} (A:C)$$

the optical density O of the solution containing the interacting species is given by

$$\frac{1}{O} = \frac{1}{\varepsilon K[A_0][B]} + \frac{1}{[A_0]} + \frac{G[C]}{K[A_0][B]}$$

where $[A_0]$ is the initial concentration of A and the other concentrations are at equilibrium. ε is the extinction coefficient at the wavelength of determination. A plot of $1/O$ vs. $[C]$ for constant $[A_0]$ and $[B]$ will yield an intercept of $O_1/(K[A_0][B]$ and a plot of these intercepts vs. $1/[B]$ gives K and ε.

Many objections have been raised concerning the validity of the simple Benesi-Hildebrand type equations (19). Some of them can be answered by invoking solvation effects as above. The validity of these equations over all range of concentrations and over different ranges of complexing strength have been discussed by many authors and various criteria given for under what circumstances the equations or their modifications may be used (19). In addition several authors have produced sophisticated statistical methods for minimizing the errors inherent in such equations. Examples of these are given by Foster (2) and Mulliken and Person (2). Nevertheless, the vast majority of workers have used the Benesi-Hildebrand without qualification and this is reflected in the sometimes conflicting values given for charge transfer complexes in the literature.

Although the equations were derived for optical absorbance changes, any physical change associated with complexing can be analysed by these equations as for example signal shifts in NMR spectroscopy.

One reason put forward to explain the failure of the Benesi-Hildebrand equation particularly in its sometimes bad predictions of ε_{CT} as compared to the values predicted by the Mulliken theory is that charge transfer can also take place between donor and acceptor without any binding, so that even when bound complexes are formed in solution there is a further contribution to the charge transfer intensity from pairs randomly in contact. Orgel and

Mulliken have examined a situation where in addition to the charge transfer tensity of the bound complexes, contact charge transfer occurs between uncomplexed donor and acceptor molecules (20). They assume that each uncomplexed acceptor molecule is in random contact with x donor molecules and the extinction coefficient of each contact is $\varepsilon_{contact}$. The measured extinction coefficient

$$\varepsilon = [1 - (q/K_c)] \, \varepsilon_{CT}$$

where ε_{CT} and K_c are the parameters for the bound complexes and

$$q = x\varepsilon_{contact}/\varepsilon_{CT}.$$

A quantum-mechanical explanation by Murrell of contact charge transfer has already been given in Section 1.3.

The charge transfer spectra seen in solutions of charge donors in oxygenated solutions almost certainly arise from contact as there is no evidence at all for binding (21). These solutions obey the Beer-Lambert law and a Benesi-Hildebrand plot extrapolates back to the origin. This is interpreted as meaning not that ε_{CT} is infinite but that no bound species are formed. Carter, Murrell and Rosch (14) believe that these results arise from solvation, not from contact charge transfer, which can give an extrapolation very close to the origin. Conversely, Slifkin and Allison (22) note that the charge transfer spectra of solutions of biomolecules show no peak absorption. The absorbance continually increases to shorter wavelength. This is explained as being due to the charge transfer transition occurring between two non-bonding levels which will not possess a maximum.

1.8. Thermodynamic Parameters

Complexes may be characterized in terms of thermodynamic parameters. Undoubtedly, these parameters are more informative than equilibrium constants. The enthalpy of dissociation ΔH° is the most commonly used parameter and is obtained from the variation of K_c with absolute temperature from the van't Hoff reaction isochore. Thus,

$$\frac{d}{dT}(\ln K_c) = \frac{-\Delta H^\circ}{RT^2} \quad \text{or} \quad \frac{\ln K_{c1}}{\ln K_{c2}} = \frac{-\Delta H^\circ}{R}\left(\frac{1}{T_1} - \frac{1}{T_2}\right).$$

This expression is only strictly correct if the association constants are expressed in inverse mole fraction, i.e. K_x. However the use of inverse moles, i.e. K_c, will introduce little error in solution.

Other parameters which can be obtained are the standard free energy ΔF° given by

$$\Delta F^\circ = -RT \ln K_c$$

and the standard entropy ΔS° given by

$$\Delta F^\circ = \Delta H^\circ - T\Delta S^\circ.$$

The enthalpy of dissociation measures the energy required to break the bond between the components in the complex and is typically of the order of a few kcal/mole. It is a measure of the binding forces. The standard free energy measures the difference in energy between the free and associated states and includes energy changes due to conformational changes and solvation effects. The entropy change is a measure of the change of order of the system, an ordered system having a lower entropy than a less ordered system. The formation of a complex will usually result in a decrease in entropy as the complexed molecules are in a more ordered state then when randomly dispersed. However where solvation occurs, it is possible for the unassociated state to be more ordered than the complexed state due to clustering of solvent molecules round the free components.

Another contribution to entropy change can come if there is a difference in conformation of either of the interacting molecules. Such changes could perhaps occur for large molecules such as proteins whose shapes are determined by weak intramolecular forces of the same strength as charge transfer forces and which could be disrupted by them. The entropy change is a useful parameter to evaluate particularly in biomolecular systems as it does indicate whether there are marked conformational or solvation effects occurring on complexing. For organic complexes $\Delta S°$ is generally proportional to $\Delta H°$ (2).

Some authors prefer to quote $\Delta F°$ values as a measure of the strength of complexing. The majority use $\Delta H°$. These two parameters are not necessarily proportional and therefore the order of complexing for groups of molecules may well be different depending on the chosen parameter. $\Delta H°$ is the parameter which is most closely related to the magnitude of the charge transfer forces and should be used for comparing complex stability.

It should be stressed that the foregoing relationships are true only for ideal solutions in inert solvents with unit activity coefficients. Such conditions are only likely to be met in carefully selected solvents and at very high dilution.

1.9. Stoichiometry of Charge Transfer Complexes

Charge transfer complexes have well defined simple stoichiometries. Although in the previous discussion we have only dealt with 1 : 1 complexes, other complexes such as 1 : 2, 2 : 3 and 1 : 1 : 1 exist in biomolecular systems. Other stoichiometries may well occur. There are various methods of determining the stoichiometries. Obviously for solid complexes, one may carry out a chemical analysis.

The best method of obtaining stoichiometries in solution is the Job plot (23). This consists of plotting some function associated with complex formation against the concentration of the two partners. The overall molar concentration is kept constant and only the fractional molarities of the components adjusted. The chosen property will reach a maximum when the molarities are in the same ratio as the stoichiometry.

This can be shown as follows. For the complex of D and A

$$mD + nA \overset{K_c}{\rightleftharpoons} (D_m:A_n)$$

$$K_c = \frac{[(D_m:A_n)]}{[D - m(D_m:A_n)]^m [A - n(D_m:A_n)]^n}$$

Differentiation of this equation shows that a maximum value of $(D_m:A_n)$ is obtained when $n[D] = m[A]$ for the condition $[A] + [D]$ is a constant.

Alternatively the optical density O is given by

$$O = K_c \varepsilon_{CT} [D]^m [A]^n, \quad \text{for } [A] \text{ and } [D] \gg [D_m A_n].$$

So that a plot of the optical density $vs.$ one component, the other being kept constant, on log paper will have a slope equal to the index in the equation.

It would therefore appear that a plot of the Benesi-Hildebrand equation if linear would show that the complex was 1 : 1. Unfortunately one of the strictures voiced against the Benesi-Hildebrand equation is that under certain conditions straight lines are apparently obtained even when the stoichiometry is not 1 : 1 (19).

Another criterion sometimes used for 1: 1 stoichiometry is the presence of isosbestic points in the spectra of mixtures in which complexing is occurring. An isosbestic point is a point of constant absorbance. In a situation where a charge transfer band overlaps to some extent an intrinsic band of a partner, there will be a point where the extinction coefficients of the two bands are the same. If therefore for every molecule of donor or acceptor removed from solution, a molecule of complex appears, at the region where the extinction coefficients are the same there should be no change in absorbance. Such isosbestic points do occur. The theory of isosbestic points has been given by Gouterman and Stevens (24). More recently an analysis has shown that isosbestic points may be observed in solutions where the solvent itself interacts with the donor and acceptor (25) so that isosbestic points may not automatically be taken to mean the presence of 1 : 1 stoichiometry.

1.10. Structure of Charge Transfer Complexes

The structure of the complexes will be determined by the intermolecular forces, the structure giving the minimum potential energy when all intermolecular forces are considered being the most probable. Considering charge transfer forces alone we note from Section 1.3 that the coefficients giving the relative weight of the dative bond wavefunction in the ground state,

$$b = -\frac{a(H_{01} - W_0 S)}{W_1 - W_0}$$

can be written as, $b \approx cS/W_1 - W_0$ where c is a constant, as $H_{01} \approx S$ and $W_0 S \ll H_{01}$. From the quantum mechanical maximum overlap principle, the minimum potential energy will occur for maximum overlap of the highest filled orbital of the donor with lowest unoccupied orbital of the acceptor, S_{DA}, where S_{DA} is given by

$$S = \sqrt{2} S_{DA}/(1 + S_{DA}^2)^{\frac{1}{4}}.$$

Thus for no overlap, $b = 0$ and no complex is formed. The contribution of the dative structure to ground state stabilization by charge transfer forces increases with the overlap integral S_{DA}.

In structural terms, this means that for a π-electron donor and acceptor, a staggered sandwich type complex is formed with the plane of the donor parallel to the plane of the acceptor†. This is only strictly true if the charge transfer forces are the only ones present. In practice, other weak forces such as van der Waals forces will be acting so that the final configuration will be that which minimizes the total potential energy due to all forces, although it will not be very different from the sandwich structure. One should also bear in mind that due to contributions from structures of the form D^+ and A^- which are slightly different in shape to the neutral forms, there will be a slight change in structure of the molecules when complexed.

The structure of complexes involving n-donors will be different. The overlap between the lone-pair orbitals on the donor with the acceptor unfilled orbital will give rise to structures where the plane of the donor, pyridine for example, will be perpendicular to the plane of the aromatic acceptor.

X-ray crystallographic studies of solid complexes will therefore differentiate between possible modes of complexing. As such measurements also give intermolecular distances, it is possible to detect the influence of charge transfer forces as these shorten the intermolecular distance as compared to other weak binding forces (9).

1.11. Chemical Substitution in Charge Donors and Acceptors

The donor and acceptor properties of molecules can be modified by chemical substitution. Substituents fall into two classes; those which are electron donating and increase the donor properties of the molecule and increase the accepting properties, and those which are electron withdrawing which increase the acceptor properties of the molecules and decrease the donor properties.

Alkyl substitution increases the donor ability while decreasing the acceptor ability of the substituent due to a flow of charge from the alkyl group to the

† A direct superposition of an aromatic donor over an aromatic acceptor will result in negative parts of the wave function overlapping positive portions of the other so that the net overlap is zero. By displacing one molecule sideways by half its width, the overlap will be an optimum.

TABLE 1.1

Some electron withdrawing and donating groups[a]

Group R	Dipole Moment[b] μ (Debyes)	Direction of Dipole R	Ionization[c] Potential (eV)
OH	1.6		8.5
NH$_2$	1.5	$\longleftarrow$ +	7.7
OCH$_3$	1.2		8.2
CH$_3$	0.3		8.8
H	0.0		9.2
Cl	1.6		9.1
CHO	2.8	+ $\longrightarrow$	9.5
SO$_3$H	3.8		—
NO$_2$	3.9		9.9

[a] After P. Sykes, 1968. "A Guidebook to Mechanism in Organic Chemistry," p. 119 Longmans, London.

[b] The dipole moment is a measure of the electron withdrawing or donating power of the substituent. Phenol at the top of the list is a good electron donor and nitrobenzene at the bottom a good electron acceptor.

[c] Watanabe, K., Nakayama, T., and Mottl, J. (1962). *J. Quant. Rad. Transfer*, **2**, 369.

rest of the molecule; a process known as hyperconjugation. Another strong electron donating group is hydroxyl. Halogens, carbonyl and indeed all negative radicals increase the acceptor ability whilst decreasing the donor ability. Thus benzene is a moderate donor and probably a poor acceptor. Table 1.1. illustrates the effect of substitution into benzene. The flow of charge into and out of the benzene ring is given by the magnitude and direction of the dipole moment. The anomolous result for aniline, is probably due to the fact that the lone-pair electrons on the nitrogen conjugate with the π-electrons of the benzene ring thus greatly increasing the overall electron negativity of the molecule whilst retaining a large proportion of the charge close to the nitrogen.

Dichlorobenzoquinone is a strong acceptor, hydroquinone is a strong donor whereas dichlorohydroxybenzoquinone is not an acceptor at all, which shows that the electron donating ability of one hydroxyl group is roughly equal to the electron withdrawing ability of two chloro groups (26).

Doubts about the role of charge transfer in complexing can be resolved by examining families of molecules containing different substituents to see whether the strength of the complexing correlates with the electron donating or withdrawing properties of the substituents.

1.12. Spectral Shifts of Donor and Acceptor Bands in Charge Transfer Complexes

One feature of many charge transfer complexes is the shifting of absorption bands of one or both of the components. The blue-shift of complexed iodine is well known (27). Several studies show that the major absorption band of chloranil at *ca.* 295 nm is red-shifted on complexing, a phenomenon particularly marked with strong donors (28). The blue shift of iodine has been explained by Mulliken (27) as arising because the partial donation of an electron into an unfilled orbital of iodine, which is in fact an antibonding orbital, increases the size of the molecule and hence its excitation energy thus shifting the absorption band to the blue. Another explanation has been put forward by the writer (28) who points out that in both cases the shift of the absorption is to the spectrum of the negative ion. The resultant spectrum is therefore due to contributions to the spectrum from both neutral and ionized forms. The fractional shift between the neutral and ionic spectra should therefore be a measure of the ratio of the coefficients of the ground state wavefunction in the Mulliken theory, $(b/a)^2$.

The amount of shift in both explanations should therefore correlate with the strength of complex formation. This has been reasonably well established for iodine complexes (29) and complexes of methylviologen with naphthoic acid and indole derivatives (30). This doesn't appear to be the case for complexes of 7,7,8,8-tetracyanoquinodimethane (TCNQ) with hexamethylbenzene, where the spectrum of TCNQ can be shifted in either direction (31). These shifts are however small when compared to iodine or chloranil shifts and might arise in part from slight perturbations caused by environmental changes on the addition of the donor to the solvent.

1.13. Infra-red Spectra of Charge Transfer Complexes

The infra-red spectra of weak complexes are similar to the superposition of the spectra of the uncomplexed components (10, 32) although for hydrocarbon-quinone complexes slight red-shifts of the carbonyl and $-C=C-$ bands of the quinone have been reported (33). These have been explained as being due to slight charge donation into an antibonding orbital of the quinone causing a slight weakening of the molecule's binding forces thus weakening the carbonyl and $-C=C-$ bonds.

Complexes involving very strong donors and acceptors appear to be almost wholly dative, i.e. ionic, in the ground state and the infra-red spectra are simply the sum of the spectra of the ions. Examples of this are the phenothiazine iodine and phenothiazine chloranil complexes (34).

Recently the writer has shown that there is an intermediate class of complexes examples of which are quinhydrone, the complex between hydro-quinone and quinone, and biological n-donors such as the amino acids (35, 36). The spectra of these complexes represent an intermediate state between those of the classes just outlined. Thus the carbonyl band of free chloranil lies at 1695 cm^{-1} (35) in the weak hydrocarbon complex at 1690 cm^{-1} (33), of the ion as observed in the phenothiazine complex at 1580 cm^{-1} (34) and the amino acid or quinhydrone complexes at 1633 cm^{-1} (35, 36). Similar orders of shifts are observed for the $-C=C-$ bands. These can all be explained as due to donation of an electron into an antibonding orbital of the acceptor reducing the stability of the molecule resulting in a weakening of the bonding. Those complexes which show marked shifts in the infra-red also show marked shifts of the acceptor spectrum in the ultra-violet and visible. In solvents of high dielectric constant many of these complexes exhibit the spectra in the visible of the acceptor anion or donor cation (37, 38).

The carbonyl band is also sensitive to hydrogen bonding (39) and therefore by no means can all shifts be attributed to charge transfer complexing. Buvet (40) has shown that under rather extreme circumstances, i.e. on complexing with bromine, carbonyl can become a charge donor which also results in a red-shift.

One property of infra-red spectroscopy is that it enables structures of molecules to be determined by the identification of absorption bands associated with particular chemical bonds. Infra-red spectroscopy has established that amino acids which are normally zwitterionic in the free state are union-ized when complexed (36). See Section 3.3.

One possible effect of complexing on infra-red spectroscopy is the alteration of the selection rules due to the lowering in symmetry of the complexed donor or acceptor compared to the uncomplexed molecules. Whilst there appears to be little experimental work done on this the topic is discussed by Foster and Mulliken and Person (2). This would allow bands which are forbidden because of symmetry considerations to be allowed in the complex.

It has been shown that for a series of related organic donors with a common acceptor, shifts of certain infra-red bands occur which can be correlated with ionization potentials (41). This might have application to biological systems.

1.14. Emission of Charge Transfer Complexes

Most charge transfer complexes appear to be non-fluorescent. Thus the addition of a donor to a fluorescent acceptor in solution, or vice-versa, causes a quenching of the fluorescence which is proportional to the association constant of the complex and can thus be utilized for measuring it.

Consider the following processes taking place during fluorescence and complexing:

1. $A + hv \rightarrow A^*$ (absorption with rate constant I.)
2. $A^* \rightarrow A + hv$ (fluorescence with rate constant $k_1[A^*]$.)
3. $A^* \rightarrow A + \text{heat}$ (deactivation with rate constant $k_2[A^*]$.)
4. $A + D \rightleftharpoons (D:A)$ (complexing with constant K_r association).

Under steady illumination conditions,

$$\frac{dA}{dt} = I - (k_1 + k_2)[A^*].$$

At the steady state,

$$I = (k_1 + k_2)[A^*].$$

The fluorescence yield of A alone in the absence of D is given by

$$F_0 = \frac{k_1 A^*}{I} = \frac{k_1}{k_1 + k_2}.$$

The fluorescence yield in the presence of D is given by,

$$F = \frac{k_1[A^*]}{I(1 + K_c[D])} = \frac{k_1}{k_1 + k_2} \cdot \frac{1}{1 + K_c[D]}$$

hence

$$\frac{F_0}{F} = 1 + K_c[D].$$

This is called the Stern-Volmer equation (42). A plot of relative fluorescence yield *vs.* quencher concentration gives the assocation constant. The variation of K_c with temperature will then give the thermodynamic parameters. Although many studies given in this text show a good agreement between the K_c of the Stern-Volmer equation and that of the Benesi-Hildebrand method, others do not. Undoubtedly the explanation is that complexing with the excited state can be different than with the ground state. On a simple molecular orbital picture, the energy required to move an electron from an excited donor to an acceptor will be decreased by the difference in energy between the highest occupied orbital and the lowest occupied orbital as compared to the unexcited donor.

Charge transfer phosphorescence has been observed in many cases (2) and is characterized by broad featureless bands. Plots of the wavelengths of the maximum of these bands *vs.* charge transfer bands for a family of related donors with a common acceptor give very good straight lines thus confirming the charge transfer origin.

1.15. Excimers and Exciplexes

In recent years it has been found that the emission of certain organic molecules changes at increasing concentration, a new broad feature emission band occurring to the red side of the fluorescence spectrum (43). This is known to be due to the formation of short-lived dimers between an excited molecule and an unexcited molecule which breaks up shortly after formation with the emission of the characteristic spectrum. When two identical molecules are involved, the excited dimer is called an excimer, dimerization between two unlike molecules is called an exciplex. It should be emphasized that this type of dimerization is only formed with the excited state and can in no way be detected with normal absorption spectrophotometry. It has been suggested that excimer formation can be explained by a charge transfer interaction (44) and there is a good relationship between the position of the excimer emission band and $I_D - E_A$ for the molecule involved†. However, calculations suggest that the excimer is bound by other interaction forces rather than by charge transfer interaction (43).

In the case of exciplexes there would appear to be more grounds for assuming that stabilization occurs through charge transfer. These exciplexes are formed between molecules of low ionization potential and high electron affinity and correlations have been found between exciplex emission and ionization potentials and electron affinities (45). The excited state, indeed the only state of the exciplex has been explained in a very similar manner to the Mulliken theory of the charge transfer complex:

$$\psi_E = a\psi_{(D, A^*)} - b\psi_{(D^+ A^-)}.$$

1.16. Excited State Spectra of Charge Transfer Complexes

Some studies have been made of the excited state spectra of these complexes. The absorption spectra of excited complexes have been looked at using flash-photolysis or excited spectrophotometry (see Chapter 2 for a description of these techniques). In all cases changes of the spectra as compared to that of the free components alone have been observed and attributed to the effect of complexing. In addition the lifetimes of these excited states are shortened relative to the lifetimes of states of the uncomplexed molecules. Briegleb and Schuster (46) have attributed these changes to the transition of an electron from the excited state of the donor to an unoccupied level of the acceptor. Slifkin and Walmsley have attributed the new spectra and lifetimes to a perturbation of the energy levels of the donor on complexing (47). At the present time there is insufficient evidence to decide which is the correct mechanism.

1.17. Characteristics of the Charge Transfer Band

In order to be able to identify the charge transfer band of a complex, it is useful to list some of the properties which have been found empirically.

† The coefficient of correlation, from the data in reference 44, is -0.94 (see Appendix at end of chapter).

In general these bands are intense, broad and featureless. Extinction coefficients can range as high as 50 000 although many of the coefficients listed in this text are as low as 500. Briegleb has pointed out that charge transfer bands are frequently assymmetrical being broader on the high frequency side (2). He has given the following empirical equation relating the frequencies at the half-width v_H and v_L to the frequency at the maximum v_{max},

$$v_H \text{ and } v_L = 2.4(v_{max} - v_L)$$

instead of

$$v_H - v_L = 2(v_{max} - v_L)$$

for a perfectly symmetrical band where v_H is the frequency on the high frequency side of the half-width.

Briegleb has also given an empirical expression of the half-width of the charge transfer transition,

$$v_{max} - v_L = 0.104 v_{max}.$$

There are several cases reported of multiple charge transfer bands (2). A probable explanation for this is the presence of close lying filled donor orbitals or unfilled acceptor orbitals, so that electron transitions take place from or to more than one orbital, each with its own charge transfer band. If the orbitals are very close together in energy, the charge transfer bands overlap, giving the appearance of a single band with an abnormally large half-width.

Fine structure is sometimes observed on the charge transfer spectra of oxygenated aromatic hydrocarbon solutions (21). This structure is the singlet-triplet spectrum of the hydrocarbon, a normally forbidden transition. Murrell has shown theoretically that among the allowed transitions of the dative structure are ones which are in fact the singlet-triplet transitions of the donor (48).

The charge transfer band can be affected by solvents. Weak charge transfer complexes which have little dative character in the ground state, show slight wavelength shifts of the charge transfer band maximum which correlate roughly with the polarity dielectric constant, or refractive index of the solvent (2).

Strong charge transfer complexes, with predominantly dative character in the ground state can dissociate into the component ions in solvents of high dielectric constant (37) so that increasing the polarity of the solvent causes the charge transfer band to be replaced by the spectra of the ions. One such type of complex is that between the pyridinium ion and a halide ion (see Chapter 8). The charge transfer band position is very sensitive to solvent. Kosower (49) has suggested that this arises from the ionizing power of the solvent and has defined a Z-value of a solvent as the position of the charge transfer band maximum of the 4-carbomethoxy-1-ethylpyridinium iodide salt in kcal/mole using the Planck relation $E = hv$ where E is the energy, h the Planck constant and v the frequency. Z-values range from 58 for chlorobenzene (50) to 94.6 for water (49). These pyridinium iodide complexes are interesting in that they are primarily ionic in the ground state and primarily non-ionic in the excited state, the absorption of light resulting in the back transfer of charge. Thus the excited states of these complexes can be represented as

$$\psi_E = a^* \psi_{(A,D)} - b^* \psi_{(A^- D^+)} \quad \text{where } a^* \gg b^*.$$

1.18. Identification of Charge Transfer Complexes

In the foregoing sections, the various properties of charge transfer complexes have been discussed. In order that charge transfer complexes can be positively identified, the following criteria should be applied. The first essential criterion is that any property associated with the presumed complex must be temperature reversible. Although this is a very obvious and necessary criterion, it does not appear to have been applied in much of the work presented in this text. An accompanying criterion is that the property should be reversible on dilution, i.e. obey the Benesi-Hildebrand equation or some equivalent.

One criterion which it is desirable to observe is the charge transfer transition. However, this may not be observed for several reasons. Firstly for donors of relatively high ionization potentials and acceptors of relatively low electron affinities, the charge transfer band would fall in the far ultraviolet region and would probably be very difficult to observe. Many of the complexes appearing here have low $K_c \varepsilon$ values so that although the charge transfer transition is in the visible region it is only observed with solid complexes (for example Fig. 7.3).

Statistical correlations between various parameters can be used for the identification of charge transfer transitions and charge transfer stabilization. As has been shown, the charge transfer transition maximum frequency should give a linear correlation with the ionization potentials of the donors and the electron affinities of the acceptors. Appreciable charge transfer stabilization of the ground state of the complex will give a correlation between the enthalpies of dissociation of the complexes and the ionization potentials of the donors or electron affinities of the acceptors. In practice, many authors use the association constants instead of the enthalpies of dissociation (cf. Section 1.7).

APPENDIX

Correlation

Two parameters x and y are linearly related by the equation

$$y = ax + b.$$

With a given set of experimental parameters x_i, y_i, the best straight line is give by

$$a = \frac{(\Sigma y_i)(\Sigma x_i^2) - (\Sigma x_i)(\Sigma x_i y_i)}{N\Sigma x_i^2 - (\Sigma x_i)^2} = \text{the slope}$$

$$and \quad b = \frac{N\Sigma x_i y_i - (\Sigma x_i)(\Sigma y_i)}{N\Sigma x_i^2 - (\Sigma x_i)^2} = \text{the intercept where there are } N \text{ pairs of observations and } y \text{ is the dependent variable.}$$

This method is known as the least-squares method as it minimizes the sum of the squares of the values of y_i from the equation of the line given by the above. The above equation assumes that there is a linear relationship between x and y. What is often required is to test the linearity between two parameters. A graph of x vs. y will hardly ever be a perfectly straight line due to experimental uncertainties or random error in the values. Correlation can however be put onto a statistical basis as follows. The scatter of the experimental points about the line obtained using the above relationships should be symmetrical about the line. The correlation is defined as the ratio of the explained variation or scatter to the total variation or scatter where the explained variation is given by $\Sigma(y_{est} - y)^2$ and the total variation by $\Sigma(y_i - y)^2$. y_{est} is the value of y given by the above equation of a straight line for the experimental value of x_i. y is the mean value of the experimental values y_i. The coefficient of correlation

$$r = \pm\sqrt{\frac{\text{explained variation}}{\text{total variation}}}$$

r varies from $+1$ for a perfectly linearly correlation to -1 for a perfectly linear negative correlation, i.e. x increasing for decreasing y. In terms of the experimental values

$$r = \frac{N\Sigma x_i y_i - (\Sigma x_i)(\Sigma y_i)}{\{[N\Sigma x_i^2 - (\Sigma x_i)^2][N\Sigma y_i^2 - (\Sigma y_i)^2]\}^{\frac{1}{2}}}.$$

A correlation of coefficient of $+0.7$ would mean that about half, i.e. $(0.7)^2$ the variation of y with x is explicable by a linear relationship between the two.

It is not sufficient to obtain a value of the correlation coefficient. It is necessary to assess how significant the result is. This can be done by statistical sampling theory. This theory enables us to decide the confidence or the probability that a correlation coefficient is different from zero. One can use "Student's" t distribution. The statistic t is defined by

$$t = \frac{r\sqrt{N-2}}{\sqrt{1-r^2}}$$

where $N-2$ are the degrees of freedom. The values of t obtained in this way can then be looked up in tables giving the confidence or probability limit of t for different degrees of freedom. There is no theoretical value which decides whether a correlation is significant or not, but the majority of authors usually take a lower confidence level of 95% as being significant.

REFERENCES

1. Mulliken, R. S. (1952). *J. Am. chem. Soc.* **74**, 811; (1952) *J. phys. Chem.* **56** 801.
2. Some general texts are:
   Briegleb, G. (1961). "Elektronen-Donator-Acceptor-Complexe." Springer-Verlag, Berlin.
   Andrews, L. J. and Keefer, R. M. (1964). "Molecular Complexes in Organic Chemistry." Holden-Day, San Francisco.
   Rose, J. (1967). "Molecular Complexes." Pergamon, Oxford.
   Foster, R. (1969). "Organic Charge-Transfer Complexes." Academic Press, London.
   Mulliken, R. S. and Person, W. B. (1969). "Molecular Complexes." Wiley, Interscience, New York.
3. Very many calculations are given in the text:
   Pullman, B. and Pullman, A. (1963). "Quantum Biochemistry." Wiley, Interscience, New York.
4. Murrell, J. N. (1958). *J. Am. chem. Soc.* **81**, 5037.
5. A representative opinion is:
   Dewar, M. J. S. and Thompson, C. C. (1966). *Tetrahed. Supp.* **7**, 97.
6. A representative view is given by:
   Kosower, E. M. (1966). "Flavins and Flavoproteins" (Ed. E. C. Slater). Elsevier, Amsterdam.
7. Mulliken, R. S. and Person, W. B. (1969). *J. Am. chem. Soc.* **91**, 3409.
8. Bent, H. A. (1968). *Chem. Rev.* **68**, 587.

9. Wallwork, S. C. (1961). *J. chem. Soc.* 494.
10. Kainer, H. and Otting, W. (1955). *Chem. Ber.* **88**, 1921.
11. Matsunaga, Y. (1966). *Nature*, **211**, 183.
12. Slifkin, M. A. and Walmsley, R. H. (1969). *Experientia*, **25**, 930.
13. Szent-Györgyi, A. (1960). *Proc. natn. Acad. Sci.* **46**, 1334.
14. Carter, S., Murrell, J. N. and Rosch, E. J. (1965). *J. chem. Soc.* 2048.
15. Rosenthal, I. (1969). *Tetrahed. Lett.* **39**, 3333.
16. Benesi, H. A. and Hildebrand, J. H. (1949). *J. Am. chem. Soc.* **71**, 2703.
17. Scott, R. L. (1956). *Rec. Trav. Chim.* **75**, 787.
18. Brooke, D. and Guttman, D. E. (1968). *J. Pharm. Sci.* **57**, 1206.
19. Deranleau, D. A. (1969). *J. Am. chem. Soc.* **91**, 4044.
    Emerslie, P. H., Foster, R., Fyfe, C. A. and Horman, I. (1965). *Tetrahed.* **21**, 2843.
    Person, W. B. (1965). *J. Am. chem. Soc.* **87**, 167
    and papers cited therein.
20. Orgel, L. E. and Mulliken, R. S. (1957). *J. Am. chem. Soc.* **79**, 4839.
21. Evans, D. F. (1954). *J. chem. Soc.* 1351.
22. Slifkin, M. A. and Allison, A. C. (1967). *Nature*, **215**, 949.
23. Job, P. (1925). *C.r. hebd. Séanc. Acad. Sci. Paris*, **180**, 928.
24. Gouterman, M. and Stevenson, P. E. (1965). *J. chem. Phys.* **37**, 2266.
25. Timberlake, C. F. and Bridle, P. (1967). *Spectrochim. Acta*, **23A**, 313.
26. Slifkin, M. A., Smith, B. M. and Walmsley, R. H. (1969). *Spectrochim. Acta*, **25A**, 1479.
27. See p. 156 to p. 161 Mulliken ref. 2.
28. Slifkin, M. A. (1963). *Nature*, **198**, 1301.
29. Mulliken, R. S. (1956). *Rec. Trav., Chim.* **75**, 845.
30. Cann, J. R. (1969). *Biochemistry*, **8**, 4036.
31. Holm, R. D., Carper, W. R. and Blancher, J. A. (1956). *J. phys. Chem.* **71**, 845.
32. Eastman, J. W., Androes, G. M. and Calvin, M. (1962). *J. chem. Phys.* **36**, 1197.
33. Slifkin, M. A. (1970). *Chem. Phys. Lett.* **7**, 195.
34. Matsunaga, Y., (1963). *Helv. Phys. Acta*, **36**, 800; (1964) *J. chem. Phys.* **41**, 1609.
35. Slifkin, M. A. and Walmsley, R. H. (1970). *Spectrochim. Acta*, **26A**, 1237.
36. Slifkin, M. A. and Walmsley, R. H. (1969). *Experientia*, **25**, 930.
37. Foster, R. and Thomson, T. J. (1962). *Trans. Farad. Soc.* **58**, 860.
38. Slifkin, M. A. (1964). *Spectrochim. Acta*, **20**, 1543.
39. Bellamy, L. J. (1968). "Advances in Infra-red Group Frequencies." Methuen, London.
40. Buvet, R. (1969). Abstr. 3rd Int. Biophys. Congress. Cambridge, USA.
41. Kagiya, T., Sumida, Y. and Inove, T. (1968). *Bull. chem. Soc. Japan*, **41**, 767.
42. Stern, O. and Volmer, M. (1919). *Physik. Z.* **20**, 183.
43. The most modern review of this topic is given in Birks, J. B. (1970). "Photophysics." Wiley-Interscience, London.
44. Slifkin, M. A. (1963). *Nature*, **200**, 766.
45. Tavares, M. A. F. (1970). *Trans. Farad. Soc.* **66**, 243.
46. Briegleb, G. and Schuster, H. (1969). *Angewandte Chemie*, **8**, 771.
47. Slifkin, M. A. and Walmsley, R. H. (1971). *Photochem. Photobiol.* **13**, 57.
48. Murrell, J. N. (1960). *Mol. Phys.* **3**, 319.
49. Kosower, E. M. (1958). *J. Am. chem. Soc.* **80**, 3253 and 3261.
50. Walling, C. and Wagner, P. J. (1964). *J. Am. chem. Soc.* **86**, 3368.

CHAPTER 2

Experimental Methods and Techniques

2.1 Spectroscopy

2.1.1. ULTRA-VIOLET AND VISIBLE ABSORPTION SPECTROSCOPY

The most common technique used in the investigation of charge transfer complexes and interactions is absorption spectroscopy in the visible and ultra-violet regions of the electromagnetic spectrum. This is not surprising as classic charge transfer complexes, i.e. in the Mulliken sense are readily recognized by the charge transfer bands which in almost all cases occur in this region. Furthermore, as absorption spectra in these regions arise from electronic transitions in the molecules, modified by the vibrations of the molecules, any changes in the electronic structure of molecules due to loss or acquisition of charge would be expected to modify the normal electronic spectra of the molecules. The wide availability of automatic recording spectro-

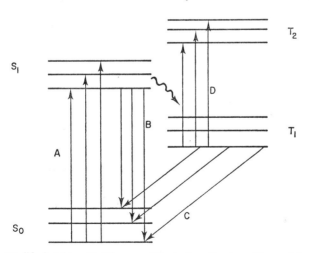

Fig. 2.1. Modified Jablonski Diagram. Illustrating the transitions giving rise to A absorption, B fluorescence, C phosphorescence and D excited state spectra, triplet-triplet transitions.

S_0 is the ground state of the molecule, S_1 is the first excited singlet state, T_1 is the first excited triplet state and T_2 is the second excited triplet state.

photometers has given impetus to work in this field. As well as conventional absorption spectroscopy, other techniques may be performed with these instruments.

2.1.1.1. *Difference Spectroscopy*

Modern absorption spectrophotometers work on the principal of comparing the difference in intensity of two identical light beams; one passing through a cuvette containing just solvent, the reference beam or channel, and the other passing through an identical cuvette containing the sample in the same solvent, the sample beam or channel. In many instances the molecules may have strong absorption in the region of interest and it is therefore difficult to

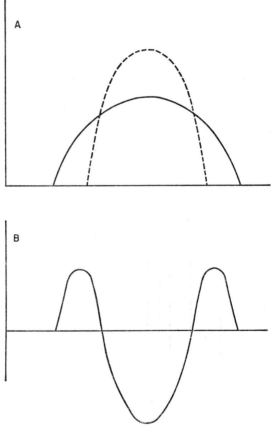

Fig. 2.2. *A* shows two imaginary absorption spectra. One of a molecule and the other on being perturbed by the presence of a second. *B* shows the difference spectra between the two with two apparent peaks.

observe slight changes due to complexing. If the changes are too small to be detected by normal spectrophotometry than difference spectroscopy may be used. There are two slightly different ways of applying difference spectroscopy.

In one approach the reference solution consists of one component in the solvent, usuually the more strongly absorbing one, and the sample channel contains the mixed solution. The spectrum obtained is then the difference in absorption between the two. As the major absorbance is effectively compensated, small changes in absorbance can be detected.

In the second approach, two cuvettes containing the two components separately are put in the reference channel while one cuvette, twice the length of the ones used in the reference beam is placed in the sample channel containing a mixture of the components. In order to balance out reflection losses from the cuvette windows, a second cuvette must be used in the sample beam containing only solvent. In this approach changes in both components can be looked for.

Difference spectra may consist of troughs and peaks. The troughs are percentage transmissions greater than 100%. These troughs or negative peaks correspond to a decrease in absorption of the sample as compared to the reference. The positive peaks may correspond to new peaks arising in the interaction. However some care has to be taken in interpreting these spectra, for if the reference species absorption band is slightly broadened and its transmission decreased due to some slight interaction possibly even due to a change in dielectric constant on the addition of the second component, then the difference spectrum will show peaks and troughs but these will not correspond to real new peaks in the sample (Fig. 2.2, facing page).

2.1.2. INFRA-RED SPECTROSCOPY

Although infra-red spectroscopy is as well-established a technique as visible spectroscopy, it has been used far less in the study of charge transfer interactions. This undoubtedly arises because what work has been carried out on organic charge transfer complexes, has in the main been done on very weak complexes with very little charge transfer in the ground state and hence very little perturbation in the infra-red. The absorption in the infra-red (2 to 16 microns) comes from vibrations of atoms or groups of atoms in the molecule. Specific absorption bands can be assigned to specific bonds between atoms. One would expect in a situation where there was any appreciable charge transfer in the ground state that this would be characterized by spectral changes in the infra-red and moreover that the site of charge transfer or acceptance could be determined. Some of the observed infra-red effects of charge transfer complexes have already been mentioned in Section 1.13.

Infra-red spectroscopy is mainly carried out with the specimens sintered in KBr discs or mulled in liquid paraffin. Spectroscopy can be carried out in solution, although for aqueous solvents, expensive cell window materials are required. In addition water is only transparent in limited regions of the infra-red. Because of the dissociation of complexes in solution and the strong infra-red absorption of water and other common solvents, it is doubtful whether solution spectroscopy can be very important in this region.

2.1.3. EMISSION SPECTROSCOPY

An alternative to absorption spectroscopy in the visible is emission, fluorescence or phosphorescence, spectroscopy. This is the reverse process to absorption. Light is emitted when a prior excited molecule reverts to the ground state. This type of spectroscopy is more difficult to perform than absorption spectroscopy. A very much higher standard of chemical purity is required. Very small amounts of impurity *ca.* 1 part in 10^6 can cause marked quenching of the emission so that little or no fluorescence or phosphorescence is detected. This quenching can come about by transfer of the excitation energy of the sample to the quencher so that the spectrum obtained is that of the impurity and not of the sample.

2.1.3.1. *Fluorescence Spectroscopy*

There are two kinds of emission spectra; fluorescence and phosphorescence. Fluorescence is observed when the excited molecules in the singlet state reverts to the ground state, the energy of excitation being released as light photons. This is a short lived process as the excited singlet molecule generally has a life time less than 10^{-7} sec. This phenomenum can be observed in solution, in KBr discs, mulls or in frozen or solid solution. Fluorescence studies yield information about molecular interactions in various ways. Most charge transfer complexes are non-fluorescent so that the removal of free molecules from solution causes a decrease in the fluorescence intensity. Even in situations where normal absorption spectrophotometry indicates little change, marked fluorescence quenching can occur.

The association constants of the complexes giving rise to quenching can be derived from the Stern-Volmer equation, namely that $I_0/I = 1 + K[Q]$ where I_0 is the fluorescence intensity of the unquenched solution, I is the intensity of the quenched solution when a molar concentration Q of quencher i.e. complexing reagent is present and K is the association constant of the complex in M^{-1}, provided that the complex is non-fluorescent and that the quencher does not alter the absorption of the solution in the region of excitation or emission to any marked degree (see p. 21).

Many molecules will interact with others in the excited state but not in the ground state so that rather than observing a ground state interaction resulting in non-fluorescent complexes, one observes the quenching of the excited state by the state itself interacting with the quencher. Usually, fluorescence spectra are the mirror image of the absorption spectra if there is no inter-action or change of configuration in the excited state. Excited state interaction can be recognized if there is a change of spectrum on the addition of the quencher. Differences between association constants obtained from absorp-tion spectra and from fluorescence have been attributed to the difference strength of interaction between the ground state and the excited state. How-ever, the limitations of the methods of evaluating association constants may also be an explanation of such discrepancies.

2.1.3.2. *Phosphorescence Spectroscopy*

Phosphorescence occurs when the initially singlet excited molecule inter-converts to a triplet excited state. There is a very low transition probability between this triplet state and the ground state and hence the lifetime may be of the order of seconds. In order that these long lived states should not be quenched by collisions or otherwise radiationlessly deactivated, it is necessary to use solid solutions to observe the phosphorescence. Normal solvents frozen to liquid nitrogen temperatures are frequently employed. Boric acid glasses which are solid at room temperature but molten at slightly elevated temperatures are also popular. A recent development is the use of solid solutions in plastics such as the polymethylacrylates. As phosphorescence is a function of the triplet state, it can provide information that no other technique other than ESR can provide. The triplet state is important in biology because of its long life and chemical reactivity. Its role in biological charge transfer is less certain. Work has been carried out on the role of the triplet state in organic charge transfer complexes *vide* 1.17.

2.1.4 EXCITATION SPECTRA

In both fluorescence and phosphorescence working one can obtain the excitation spectrum of the sample. The light output at a fixed wavelength is plotted as a function of the wavelength of the exciting light. The excitation spectrum obtained is therefore that of the absorbing species. This could be important in mixtures when the emission is that of a complex not of an individual species.

2.1.5. FLASH PHOTOLYSIS AND EXCITED STATE SPECTROSCOPY

An alternative approach to the examination of excited states is to look at the higher energy of the states rather than their emission (Fig. 2.1). In flash photolysis, the first excited state of the molecule is populated by an intense flash of light which is followed very quickly by a second intense light flash. This second flash is dispersed and then detected usually photographically. The resultant spectrum is that of the excited state and the ground state sample (1).

A variant of this method is to excite the sample with a relatively low intensity light beam chopped at some definite frequency, whilst a constant intensity white light beam passing through the sample is monochromated and detected by a photomultiplier. Any modulation on the constant intensity beam at the chopping frequency arises from absorption by the excited state of the sample (2). The modulation is detected by a phase-sensitive detector.

In both techniques the lifetime of the excited state can be measured; in flash photolysis by directly observing the decay of absorption after the second flash and in excited state spectroscopy by observing the phase shift between the exciting beam and the modulation on the monitoring beam. Because of the marked quenching by oxygen of the excited state, rigorous deoxygenation of solutions is required.

2.1.6. SPECTROPOLARIMETRY

Many biomolecules are optical active, because of their molecular assymmetry, i.e. they rotate the plane of polarized light passing through their solutions. In regions where they are absorbent, the solutions exhibit circular dichroism. The optical rotatory power of the solutions becomes anomalous in this region, passing through a minimum at the low wavelength edge of the absorption band and reaching a maximum near the centre of the absorption band. As the rotation within the absorption band is different for left-hand circularly polarized light as from right-hand circularly polarized light, a linearly polarized beam is converted to an elliptically polarized beam, the ellipticity reaching a maximum within the band. This is called the Cotton effect. As the absorption of biomolecules is on the whole in the ultra-violet, spectrophotometric techniques must be used.

It has been found that circular dichroism and optical rotation can be associated with charge transfer bands of both inter- and intramolecular charge transfer complexes which have centres of configurational assymmetry (3).

2.1.7. SPECTROSCOPY OF SOLID COMPLEXES

Weak complexes can be highly dissociated in solution and it may be necessary to work with solid complexes. Most of the techniques described in Section 2.1 can be used. The complex can either be mulled in liquid paraffin between glass or quartz plates or sintered in KBr discs. However absorption spectra in the visible and ultra-violet are likely to be very erroneous due to scattering from the solid particles.† An alternative technique is to look at the reflectance spectra of the solids. This can give good qualitative results and

---

† The writer has found however that certain solid biomolecules and complexes in sufficiently low concentration in KBr, *ca.* 5 parts in 200 by weight appear to form solid solutions with no apparent light scatter and with resolution as good as solution spectroscopy.

enable well-resolved spectra to be obtained. The nature of the surface of the sample will have an effect on the reflectance and there will be a low-level background of scattered light, the intensity depending on the wavelength, the particle size and physical condition of the sample.

2.1.8. MICROWAVE AND RADIOFREQUENCY SPECTROSCOPY

2.1.8.1. *Electron Spin Resonance (ESR)*

The most direct method of observing the uncoupling of an electron from a molecule, is to observe the ESR of the molecule.

An electron has associated with it an intrinsic spin with quantum number $S = -\frac{1}{2}$. In the presence of an external magnetic field this electron can align itself parallel or antiparallel to the field direction. The energy of the electron in the two states is $\mp \frac{1}{2}g\beta_0 H$, where H is the magnetic field, $g\beta_0$ is the inverse gyromagnetic ratio with β_0 being the Bohr magneton ($eh/4mc$) and $g = 2\cdot0023$ for free spin. As the population of the two states of the free electron is proportional to $e^{-\frac{1}{2}g\beta_0 H}$ and $e^{+\frac{1}{2}g\beta_0 H}$, the vast number of electrons will lie in the lower energy state. When energy equal to the difference between the states, $g\beta_0 H$ is fed into such a system it will be absorbed promoting electrons from the lower to the higher energy. This forms the basis of instruments used to detect free or uncoupled electrons.

In most commonly used ESR machines, an external magnetic field of *ca.* 3000 gauss is used so that the energy $g\beta_0 H$ corresponds to electromagnetic radiation of 3 cm wavelength. Such microwave radiation can only be propagated down waveguides of fixed dimensions. In practice therefore, the sample is placed in the cavity of the waveguide and microwaves of a fixed frequency fed down the guide through the sample. A small alternating magnetic field is superimposed on the main magnetic field and absorption of the microwaves looked for.

In addition to the external magnetic field, the electron will also "see" internal magnetic fields due primarily to nearby protons, so that several peaks may be seen.

The normal way of presenting signals in ESR apparatuses is as the first derivative of the observed signal. A single absorption will be shown as a positive peak followed by a negative going peak in the first derivative. The derivative curve passes through zero where the signal goes through a maximum.

The signal can be characterized by three parameters, the g value, obtained by measuring the H for which the peak is obtained, the area under the curve and the half-width.

The g factor of $2\cdot0023$ only applies to a free electron. When coupling of the electron spin to its orbital angular momentum occurs then a slightly different *g* factor is obtained. This is what one would expect with charge transfer complexes where the donated electron is not completely free from the donor.

The area under the curve gives the number of free electrons in the sample.

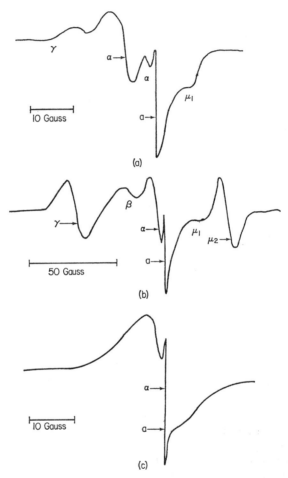

Fig. 2.3. ESR spectra of some solid charge transfer complexes. Reproduced with permission from Fig. 2, ref. 6.

Normally one would calibrate the instrument with the signal of a standard sample with a known number of electron spins.

The half-width of the signal is proportional to the inverse half-life of the electron. Interaction between electrons can lead to broadening of the signal. In a charge transfer complex where there still exists coupling between the donated electron and its (spin-paired) partner from the highest filled molecular orbital this will result in broadening. One would not therefore expect to detect a signal from a very weak complex.

The detection of ESR signals in solutions of high dielectric constant is very difficult due to strong absorption of the microwaves. This can be overcome to some extent by using very small samples and using a cavity in which the electric field vector is aligned along the length. This will minimize absorption which is an electric effect not a magnetic one. Alternatively by working with smaller magnetic fields, the energy required to promote the electron is shifted into the VHF region. VHF electromagnetic radiation is far less absorbed by polar solvent than are microwaves. As VHF will propagate down conductors, no waveguides are required and one can operate at fixed magnetic fields and tune the VHF waves through resonance. Normal radiocommunication techniques and instruments can be used. The method is also very cheap but has much poorer resolution than working with microwaves.

The situation regarding charge transfer complexes is confused. Signals have been detected from many solid charge transfer complexes. In many cases these signals have varied markedly with the mode of preparation of the sample and the presence or absence of light. In some cases the signals are thought to arise from impurities, photoionization or lattice defects in the crystalline sample. An account of ESR of organic charge transfer complexes has been given by Guttman and Lyons (4) and Foster (5).

In general the signals observed from solid charge transfer complexes are broad and structureless. One study however has revealed the presence of two peaks, a broad peak assigned to an oxidation-reduction process, i.e. a charge transfer, and a much sharper peak assigned to a donor acceptor interaction (6) (Fig. 2.3, facing page).

2.1.8.2. *Nuclear Magnetic Resonance (NMR)*

Protons possess intrinsic magnetic moments μ_p which can be aligned with or against magnetic fields in a similar manner to the electron. As the mass of a proton is so much greater than that of the electron, the magnetic moment is consequently much less than that of the electron. Consequently for an external magnetic field of 2000 gauss, an energy equivalent to a radiofrequency of about 8·5 Mhz is required to promote the proton from the low energy state of $-\mu_p H$ to $+\mu_p H$ where $\mu_p = eh/4\,Mc$. Protons are affected by the magnetic fields of neighbouring nucleons, thus giving information about the immediate environment. The presence of an unpaired electron arising from a charge transfer process will cause shifts of observed signals. However similar shifts are observed from other weak complexes such as hydrogen bonded complexes and complexes bound by dipole-induced dipole forces. NMR therefore will not distinguish between different kinds of complexes. One advantage of NMR spectroscopy is that the observed lines are so well resolved that it is

quite feasible to measure the shifts of several complexes present simultaneously together in one solution (7). Plots of shifts *vs.* concentration enables the
association constants to be measured using the Benesi-Hildebrand equation
but utilizing signal shift instead of optical density. The variation of the
association constants, measured this way with temperature enables the
thermodynamic parameters to be evaluated. At the present there appears to
be no fully successful explanation to explain the origin and magnitude of
observed shifts. NMR spectra of organic charge transfer complexes have been
reviewed by Foster (5).

2.1.8.3. *Nuclear Quadrupole Resonance (NQR)*

Nuclei which are not perfectly symmetrical will possess quadruple moments.
These moments interact with the electrical field gradients present at the
nucleus and will thus exhibit resonance on absorbing energy of the appropriate
magnitude. The character of the bonds between the nuclei has a marked
effect on the local electric field gradient and it is expected that in a situation
where there was any appreciable charge transfer in the ground state this
would be detected by NQR spectroscopy. A study of the NQR of ^{35}Cl, an
assymetric nucleus, in solid hexamethylbenzene chloranil complexes lead to
the conclusion that charge transfer is less than 5–10% in the ground state (8).

2.2. Polarography

Polarography is an electrochemical technique which enables oxidation and
reduction processes to be observed. A pair of electrodes, the cathode being a
dropping-mercury electrode, are placed in a non-aqueous solution of the
sample. An applied voltage is varied and a plot of current *vs.* voltage obtained.
Such plots are not linear but show steps when any of the molecules present

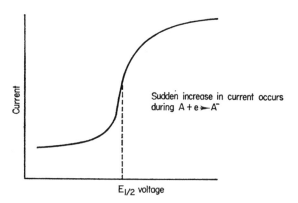

Fig. 2.4. Typical plot of polarographic experiment showing the half-wave potential $E_{\frac{1}{2}}$.

undergo an oxidation or reduction at the electrodes. The potentials at which these steps occur can be related to the redox potentials of the molecules.

Peover has analysed a system containing charge donor and acceptor which can combine forming 1 : 1 charge transfer complexes (9).

The following are the important interactions:

$$A+D \overset{K}{\rightleftharpoons} (A:D) \quad \text{and} \quad A+e \rightleftharpoons A^-,$$

where e is an electron coming from the cathode to reduce the acceptor. The difference between the half-wave potential $E_{\frac{1}{2}}$ in a system containing the acceptor alone and that with donor added is given by

$$\text{antilog}_{10} \left(\frac{0 \cdot 4343F}{RT} \Delta E_{\frac{1}{2}} + \log \frac{I_s}{I_c} \right) = 1 + \frac{K[D]\gamma_D\gamma_A}{\Sigma_i \gamma_{DA}} \text{ for } D \gg A$$

where R, T, and F are the usual thermodynamic quantities, γ terms are activity coefficients, I_c is the diffusion current constant in the mixed system and is proportional to the root of the experimental diffusion coefficient. I_s is a similar term for the solvent containing acceptor only.

With these relationships, Peover has obtained good agreement between association constants obtained from optical measurements and those from polarography.

If higher order complexes of the form

$$(D:A)+D \overset{K_2}{\rightleftharpoons} (D_2:A), \qquad (D_{N-1}:A)+D \overset{K_N}{\rightleftharpoons} (D_N:A)$$

are present, then the previous equation is modified by the addition of terms to the right-hand-side, of the form

$$+ \frac{K_2[D]^2\gamma_D^2\gamma_A}{\underset{i}{\Sigma}\gamma_{D_2A}} + \text{etc.}$$

Polarographic studies of the organic complex pyrene : tetracyanoquinodimethane indicate the presence of both 1 : 1 and 2 : 1 complexes, a result not indicated in the spectrophotometric determination of the association constant (9).

2.3. Solubility and Partition Methods

It is possible to measure association constants by measuring the increase in solubility of a compound on the addition of another with which it forms a complex. Alternatively, the partition of a compound between two different solvents in the presence of the complexing regent can be measured. The evaluation of association constants from such data is rather complicated. One requires to know the effect of the two compounds on each other in the absence of the charge transfer interaction, which is very difficult to assess. A detailed analysis of partition and solubility methods has been given by Hayman (10). The apparent association constant $K_{x\,app}$ is given by:

$$K_{x\,app} = \lim_{X_A, X_B \to 0} \frac{X_A - X'_A}{X_A X_B}$$

where X_A is the apparent mole fraction of component A in the solvent plus component B, and X'_A is the apparent mole fraction of A in solvent alone. The true association constant K_x is given by:

$$K_x = K_{x\,app} + \left(\frac{\partial \gamma_A}{\partial \gamma_B}\right)_0$$

where the terms are the rational activity coefficients of the components in dilute solution in each others presence. One advantage of this method is that the presence of higher order complexes of the type $(A_2 : B)$ etc. has no effect on K_x which is therefore the true association constant for the 1 : 1 complex $(A : B)$. The obvious disadvantage of this method is the difficulty of obtaining the activity coefficients.

A possible partition method which has been applied to organic systems is gas chromatography. This is a technique for separating out different compounds by vaporizing them and carrying them in a gas stream through a liquid. The compounds distribute themselves between the gas and the liquid so that they emerge from the apparatus at different times depending on their partition functions between the gas and liquid. It has been shown that the chromatography of electron donors, using an electron acceptor as the stationary, liquid phase, results in the retention of the donor in the liquid phase being proportional to the association constants of the complexes formed between the donors and acceptor in solution (11).

2.4. Semiconductivity

Semiconductors are materials whose specific conductivity x_0 is governed by the relationship:

$$x_{(T)} = x_0 \left[\exp \frac{(-E)}{2kT}.\right.$$

Thus a semiconductor is normally defined in terms of two parameters, the conductivity x_0 and the activation energy E. In measuring the semiconductivity of biological molecules, one is beset with two problems. In the vast majority of biomolecules, solid samples contain water and there will be a contribution to the conductivity from the protons in this water of hydration. The physical state of samples make a marked difference to measured values, due to intergranular capacitance and resistance. Powdered samples show a marked dependance of conductivity on applied pressure. Conductivity values measured by DC methods can differ considerably from those obtained by AC methods.

Rosenberg *et al.* (12), have derived an equation which is independent of the physical state of the specimen or of the amount of water of hydration *viz.*

$$x_{(T)} = x_0' \exp \frac{E}{2kT_0} \exp \frac{-E}{2kT}$$

The material is characterized by an additional parameter, the characteristic temperature T_0. Plots of x *vs.* $1/T$ for different states of the compound in question, all extrapolate back to T_0 and x_0 and x_0' are related by:

$$x_0 = x_0' \exp \frac{E}{2kT_0}.$$

A quite different approach is that of Huggins and Sharbough (13) who consider the apparent semiconductivity of dry crystalline samples to be composed of the intrinsic conductivity of the sample plus inter-granular and electrode resistance and capacitance. These artefacts can be resolved by using AC methods. Plots of $x_{(T)}$ *vs.* frequency and capacitance *vs.* frequency enable these factors to be sorted out by normal AC network analysis.

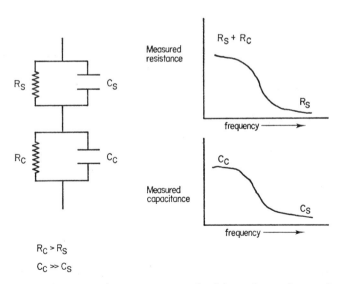

Fig. 2.5. Equivalent network to represent conductivity and capacitance of powdered specimens. R_c = integranular resistance C_c = intergranular capacitance, R_s = resistance of specimen, C_s = capacitance of specimen. After ref. 13.

The semiconductivity of organic charge transfer complexes is greatly different to the semiconductivities of the individual components. The complexes have conductivities which are several orders of magnitude greater than those of the partners. There is always a concomitant decrease in the activation energy. Various references will be made in the text to the semiconductivity of complexes of biomolecules in which charge transfer complexing has been inferred from these marked changes in conductivity.

These conductivity changes on complexing are explained by the removal of the electron from the donor to the acceptor leaving behind a "hole" into which other electrons may flow. The decrease in activation energy is explained as less energy is required to promote the electron from the donor to the acceptor than is required to promote it into the conduction band of the donor.

2.5. Conductimetric Titration

The formation of charge transfer complexes in solution of high dielectric constant may be followed by dissociation into free ions which will cause a rise in the conductivity of the solution. The conductivity will reach a maximum when the donor and acceptor are present in the stoichiometry required for the complex. The formation of 1 : 1 complexes between various molecules of biological interest have been established by this technique (14).

2.6. Mixed Fusion Analysis

This rather novel method can be used to identify the formation of complexes. Mixed fusions are prepared of possible donors and acceptors. The mixing zone is observed during solidification and any colour changes noted during the process. Other indications of complexing are the formation of eutectics, i.e. the replacement of the two melting points of the individual components with a single melting point characteristic of the mixed fusion. The stoichiometries of the complexes can be adduced from the melting point data. This method will not have wide application to biomolecules due to their instability at high temperature. It has been used by Laskowski (15) to study complex formation between menadione (vitamin K_3) and polycyclic aromatic hydrocarbons (see Section 9.6.1).

2.7. Calorimetry

A direct method of measuring the parameters of complexes is calorimetry in which the heat of mixing the two components is measured. The heat of formation and the association constant of the complex is obtained as follows:

$$\Delta H°[(A:D)] = \frac{10^3 \Delta H^m}{V}$$

where ΔH° is the heat of formation of 1 mole of complex, ΔH^m is the measured heat of formation, V is the volume of solution in litres and $[(A:D)]$ is the concentration of the complex. The association constant K_c is given by

$$K_c = \frac{[(A:D)]}{[A_0-(A:D)][D_0-(A:D)]}$$

and eliminating between these two equations gives

$$\frac{1}{K_c} = \frac{10^3 \Delta H^m}{V \Delta H^\circ} + \frac{[A_0][D_0] V \Delta H^\circ}{10^3 \Delta H^m} - ([A_0]+[D_0]) \qquad \text{ref. 16.}$$

This can be solved in the following manner. A series of arbitrary values are assumed for ΔH° and plotted against K_c obtained from the equation. The curves thus obtained intersect at the correct values of K_c and ΔH° (17).

2.8. Measurement of Dipole Moments

In the discussion of charge transfer complexes (1.3), it was shown how charge transfer complex formation gave rise to dipole moments which are directly related to the coefficients of the wavefunctions and the overlap integral S. Dipole moments of several organic systems have been measured in solution and the coefficients a and b obtained (18).

The dipole moment can be obtained from the dielectric constants of the system in a non-polar solvent. The dielectric constants are measured by finding the ratio of the capacitance of a condenser when filled with the sample and when empty using an AC bridge method.

The dipole moment μ is related to the dielectric constant E by the Clausius-Mosotti equation:

$$\frac{M}{\rho} \cdot \frac{E-1}{E+2} = P = P_3(f_3-\Delta f) + P_4(f_4-\Delta f) + P_2 \Delta f + P_1 f_1.$$

Subscripts 1, 2, 3 and 4 refer to solvent, complex and components respectively, f is the initial mole fraction, Δf is the mole fraction of complex, ρ is the density of the solution, M is the mean molecular weight

i.e. $$\frac{f_1 M_1 + f_3 M_3 + f_4 M_4}{1 - \Delta f}$$

and P is the polarization given by $P_i = 4/3\,\pi N \alpha_i$ for non-polar molecules and $P_2 = 4/3\,\pi N(\alpha_3 + \alpha_4 + \mu^2/3kT)$, N being Avogadro's number and k, Boltzman's constant. P_1, P_3 and P_4 can be determined separately by separate experiments and hence $\mu = 0{\cdot}0128\,(P_2-P_3-P_4)^{\frac{1}{2}} T^{\frac{1}{2}}$.

Foster (5) has pointed out that plots of polarization vs. weight fraction w enable the association constant K_c to be evaluated as

$$\frac{\Delta P}{(P_3^*-P_3)} = 1 + \frac{M_4^\dagger}{K_c w_4 \rho},$$

where $P_3{}^*$ is the apparent polarization constant assuming no complex formation, i.e.

$$= \frac{P - f_4 P_4 - f_1 P_1}{f_3}$$

and $P_3{}^*$ and P_3 are the values extrapolated to infinite dilution. $\Delta P = P_2 - P_3 - P_4$ for 1 mole of complex formed. A plot of $(P_3{}^* - P_3)^{-1}$ vs. $w_4 \rho$ is a straight line with an intercept of ΔP^{-1} and a gradient of $M_4 / K_c \Delta P$.

2.9. Magnetic Measurements

The Mulliken theory of charge transfer complexes predicts that the complex in the grounds state should have a permanent magnetic moment arising from from the uncoupling of an electron by donation. Hence the diamagnetic susceptibility of the complex should be somewhat less than that of the sum of the diamagnetic susceptibilities of the two free components‡. Many measurements confirm this decrease of diamagnetism on complexing (19) although there are cases of increasing diamagnetism (20). One possible explanation of this is if there are appreciable dispersion forces in the no-bond function then as shown by Dewar and Thompson (21) these forces can be represented by the valence bond model as arising from contributions to the structure of locally excited structures such as D^*,A and D,A^*. The electrons in the excited states are in larger orbitals so that they give an increased contribution to the diamagnetism. Alternative, these small changes in magnetic susceptibility could lie within the experimental uncertainty.

Diamagnetic susceptibility measurements are usually made by weighing the samples in the presence of and in the absence of a strong applied magnetic field. The effect of the application of a magnetic field to a diamagnetic sample is to cause a force at right angles to the direction of the field. For a field in the y direction, the force at right angles to it F_x is given by

$$F_x = \tfrac{1}{2}(X_1 - X_2)\mu_0 v \frac{dH^2}{dx}$$

† Foster's use of M_3 here would appear to be a misprint.
‡ Molecules with no permanent magnetic dipole moment, i.e. those containing paired electrons are diamagnetic. In the presence of an external magnetic field the electrons precess in such a way as to produce an opposing magnetic field. Thus the magnetic susceptibility is negative. The magnetic susceptibility is proportional to the radii of the filled electronic orbitals. The majority of molecules in this text are diamagnetic. Molecules or complexes which possess permanent magnetic moments are paramagnetic. In the presence of an external magnetic field, the dipole aligns itself with the applied field and the magnetic susceptibility is positive. Hence an increase in paramagnetism will be manifested as a decrease in diamagnetism. The diamagnetic susceptibility has the form $X_p = -N_p{}^2/6mc^2 \Sigma^{-2}$ and the paramagnetic susceptibility $X_D = m_\mu{}^2 N/3kT$ where m_μ is the magnetic moment of the molecule.

where H is the applied field, X_1 and X_2 are the magnetic susceptibilities of the sample and the medium in which it is immersed, usually the atmosphere and v is the volume.

There are now available very sensitive commercially made apparati for measuring magnetic susceptibility and perhaps this method could be utilized in a study of biomolecular complexes.

There is one very important difference between diamagnetic and para-magnetic susceptibility. The former is temperature independent whereas the latter varies as the inverse of the absolute temperature. Consequently, a study of the magnetic susceptibility as a function of the absolute temperature should enable the paramagnetic susceptibility to be measured. To the writer's knowledge this has not been attempted for charge transfer complexes but it would appear to be a worthwhile experiment.

2.10. Other Methods

Other physical properties which might be changed by charge transfer interaction are; surface tension, refractive index, viscosity and vapour pressure. Any of these changes could be utilized in obtaining association constants of the complexes. Rose gives several references to these methods in his monograph (22).

2.11. Kinetic Methods

Up to this point the discussion has been concerned with evaluating the parameters of complexes at the equilibrium state. It is quite feasible to work out both forward and back rate constants as well as equilibrium constants by observing the rate of growth of a complex. The equilibrium constant is the ratio of the forward and back rate constants.

For the reaction,

$$A + D \underset{k'}{\overset{k}{\rightleftharpoons}} X$$

where X is the complex of A and D, the concentration at equilibrium is given by

$$k([A - X_0])([D - X_0]) = k'[X_0]$$

where the subscript 0 represents equilibrium concentrations. In general

$$\frac{d[X]}{dt} = k([A - X])([D - X]) - k'[X]$$

Then for $[D] \gg [A]$ which implies $[D] \gg [X_0]$ and $[X]$, by eliminating k'

between the two above equations,

$$\frac{d[X]}{dt} = k[A-X][D]\left(1 - \frac{[X]}{[X_0]}\right) \quad \text{or} \quad \frac{d[X]}{dt} = k[A][D]\left(1 - \frac{[X]}{[X_0]}\right)$$

for very small K_c or $D \sim A$. A plot of the concentration of the complex (which may be the absorbance $\times \varepsilon \times$ the pathlength) vs. time will give a curve obeying the above equation. The parameters of this curve can then be solved by simulating the reaction on a computer and comparing computed

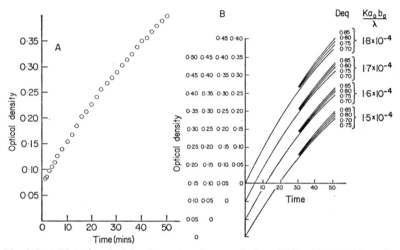

Fig. 2.6. A. Plot of optical density vs time from a solution of 4.7×10^{-5} M chloranil and 1×10^{-2} M proline in 50% aqueous ethanol, buffered to pH 9. B. Plots of optical density vs time for different values of concentrations and reaction rates as obtained from the computer. Reproduced with permission from Figs. 1 and 3 of ref. 23.

curves with experimental curves. Such a solution gives k and X_0 and hence k' is given by the equilibrium equation above. Finally K is obtained from $K = k/k'$.

This method has been described in detail for the proline-chloranil complex (23) (see Section 3.4 and Fig 2.6).

Two methods may be used for obtaining the change of complex concentration. If the interaction rate is slow, the growth of the complex may be monitored on mixing. For fast interaction rates more sophisticated methods are called for. The equilibrium condition can be disturbed by sending a fast high temperature pulse into the system which causes a momentary dissociation of the complex. The reassociation is then observed. By using repetitive pulses and averaging over many cycles with a computer of average transients, it should be possible to obtain very accurate reaction curves.

An alternative method of evaluating association constants is to observe the modification of chemical reaction rates when one or more of the reactants are removed by complex formation with an added compound. Various references will be made in the text to such experiments. Similarly association constants can be obtained by adding a donor say to an already formed charge transfer complex and observing the decrease in the pre-existing complex.

2.12. Selection of Solvents

The choice of solvent is very important. For most biochemicals, water is the obvious solvent and is indeed the ideal solvent being highly transparent and very cheap. In many systems it is necessary to control the pH. Simple acids and alkalis are to be preferred at all times. The introduction of buffers can cause many complications. Most common buffers absorb strongly below about 300 nm and therefore impose a severe limitation in the spectral region available for investigation. Particularly so as many biomolecules and their complexes only absorb below 300 nm. Many buffers themselves can take part in charge transfer interactions and one can never be sure that the observed reactions are not being modified or being solely determined by the buffers. In the writer's experience, phosphate buffers are relatively inert whereas citric acid buffers tend to strongly inhibit charge transfer interaction.

Difficulties occur in trying to examine the interaction between polar and non-polar molecules. There is the choice of using aprotic solvents such as dimethyl sulphoxide or dimethyl formamide or using mixed solvent systems such as ethanol/water or acetone/water. The aprotic solvents are very good solvents but suffer from two disadvantages; they are relatively opaque in the ultra-violet below about 280 nm and even worse are chemically active and are themselves good charge donors. Ethanol/water mixtures are very transparent and relatively inert. They are poorish solvents both for polar and non-polar compounds. Acetone/water is rather less transparent in the far ultra-violet but is a better solvent. Both ethanol and acetone are charge donors albeit rather weak ones (24, 25, 26).

2.13 Preparation of Solid Complexes

One of the disadvantages of working with solutions of weak complexes is that they are highly dissociated. It is therefore not possible to obtain high concentrations of complexes unless one of the components is present in great excess. This means that one has to separate the properties of the complex from that of the component.

An obvious way round this is to prepare solid complexes. This can be a difficult task. The most simple way which can occasionally be used is to grind the two components together in a pestle and mortar.

Other methods which have worked in particular cases are the evaporation of solutions of the components in the correct stoichiometry or the freeze drying of such solutions. The writer has managed to prepare solid complexes of some hydrocarbon quinone complexes and some protein chloranil complexes by precipitation from solution; in the first case by adding water to solutions in acetone and in the second case by adding NaOH to solutions in 50 % aqueous ethanol. Solid protein chloranil complexes have also been obtained by freezing solutions in 50 % ethanol whereupon the complexes came down. Solid complexes have been prepared by mixed fusions of the components which solidify out as the charge transfer complex.

APPENDIX

Spectroscopic Units

Wavelengths (λ)

Wavelengths are expressed in either angstroms (Å), millimicrons (mμ) or nanometres (nm), nm being the preferred unit in the MKS system. The relationship between them is

$$10 \text{ Å} = 1 \text{ m}\mu = 1 \text{ nm} = 10^{-9} \text{ metres.}$$

In the infra-red the micron (μ) is the unit in use

$$1\mu = 10^{-4} \text{ cm} = 10^{-6} \text{ metres.}$$

Wavenumbers (v)

Frequently reciprocal wavelengths called wavenumbers are used in the presentation of spectra. The wavenumber is occasionally and erroneously called the frequency. The unit of wavenumbers is the inverse centimetre (cm^{-1}) often called the Kayser (K) so that in the visible and ultra-violet regions of the electromagnetic spectrum the unit used is the kiloKeyser (kK).

Frequency (f)

A unit related to the wavenumber seen only in very old literature is the true frequency expressed in cycles per second (cps), modern usage would be the Hertz (Hz), and related to the wavelength of wavenumber by the wave relationship $f = c/\lambda = cv$ where c is the velocity of light, nominally 3.10^8 m.

Energy (E)

Related to the above units are the units of energy the kilocalorie per mole (kcal/mole) and the electron volt (eV) by the Planck relationship; $E = hf$ where h is the Planck constant $= 6 \cdot 6\,252 \times 10^{-27}$ erg sec.

$$1 \text{ eV} = 23 \text{ kcal/mole} \equiv 1239 \text{ nm} \equiv 8 \cdot 1 \text{ kK.}$$

Absorbance and Percentage Transmission

If I_0 is the intensity of a beam of light passing through an absorbing sample and I is the intensity of the emergent beam, the percentage transmission $\%T$ is given by:

$$\%T = 100(I/I_0).$$

The absorbance (A), absorbancy or optical density is given by

$$A = \log(I_0/I) = 2 - \log \%T.$$

Absorbance is usually to be preferred in the presentation of spectra as absorbances are additive by virtue of the Beer-Lambert law *viz*:

$$I = I_0 e^{-\varepsilon cl}$$ where l is the path length in cm, c is the concentration

in moles per litre (M) and ε is the molecular extinction coefficient of the absorbing medium at the wavelength of measurement. It is quite common to express spectra in terms of ε which as can be seen from the Beer-Lambert law is directly proportional to the absorbance, and has the advantage of being a constant for the particular molecule or complex being studied as it is independent of either path length or concentration.

REFERENCES

1. Norrish, R. G. W. and Porter, G. (1949). *Nature*, **164**, 658.
2. Slifkin, M. A. and Walmsley, R. H. (1970). *J. Phys. E*, **3**, 160.
3. Carrion, J. P., Donzel, B., Deranleau, D. A., Esko, K., Moser, P. and Schwyzer, R. (1967). *Helv. Chim. Acta*, **51**, 459.
4. Guttman, F. and Lyons, L. E. (1967). "Organic Semiconductors." Wiley, New York.
5. Foster, R. (1969). "Organic Charge-transfer Complexes." Academic Press, London.
6. LuValle, J., Leiffer, A., Koral, M. and Collins, M. (1963). *J. phys. Chem.* **67**, 2635.
7. Foster, R. and Fyfe, C. A. (1967). *J. chem. Soc.* 213.
8. Douglass, D. C. (1960). *J. chem. Phys.* **32**, 1882.
9. Peover, M. E. (1964). *Trans. Farad. Soc.* **60**, 417.
10. Hayman, H. J. G. (1962). *J. chem. Phys.* **37**, 2290.
11. Cooper, A. R., Crowne, C. W. P. and Farrell, P. G. (1967). *J. Chromat.* **29**, 1; (1967) *Trans. Farad. Soc.* **63**, 447.
12. Rosenberg, B., Bhowmik, B. B., Harder, H. C. and Postow, E. (1968). *J. chem. Phys.* **49**, 4108.
13. Huggins, C. M. and Sharbough, A. H. (1963). *J. chem. Phys.* **38**, 393.
14. Gutmann, F. and Keyzer, H. (1966). *Electrochim. Acta*, **11**, 555.
15. Laskowski, D. E. (1960). *Anal. Chem.* **32**, 1171.
16. Epley, T. D. and Drago, R. S. (1967). *J. Am. chem. Soc.* **89**, 5770.
17. Rose, N. J. and Drago, R. S. (1959). *J. Am. chem. Soc.* **81**, 6138, 6141.

18. Kobinata, S. and Nagakura, S. (1966). *J. Am. chem. Soc.* **88,** 3905 and references therein.
19. Bhatnagar, S. S., Verma, M. R. and Kapur, P. L. (1934). *Ind. J. Phys.* **9,** 131.
20. Sahney, R. C., Aggarwal, S. L. and Singh, M. (1946). *J. Ind. chem. Soc.* **23,** 335.
21. Dewar, M. J. S. and Thompson, C. C. (1966). *Tetrahed. Supp.* **7,** 97.
22. Rose, J. (1967). "Molecular Complexes." Pergamon Press, Oxford.
23. Crossley, T. R. and Slifkin, M. A. (1967). *Educ. Chem.* **4,** 280.
24. Middaugh, R. L., Drago, R. S. and Nedzielski, R. J. (1964). *J. Am. chem. Soc.* **86,** 388.
25. Maine, P. A. D. (1965). *J. chem. Phys.* **26,** 1192.
26. Slifkin, M. A. (1965). *Spectrochim. Acta,* **21,** 1391.

CHAPTER 3

Amino Acids and Proteins

3.1. Introduction

Amino acids and proteins are very abundant in nature and are of great biological importance. Notwithstanding, very little work appears to have been done on these molecules with the exception of tryptophan and tyrosine.

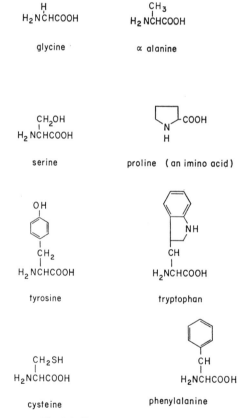

Fig. 3.1. Some common amino acids.

This undoubtedly arises from the generally accepted if perhaps not acknowledged view, that only charge transfer interactions involving π-electrons are likely to be of biological importance.

There are about 24 naturally occurring amino acids, the vast majority of which do not possess π-electrons. However the amino acids do possess an amino group ($-NH_2$) which has lone-pair electrons located on the nitrogen.

3.2. Oxygen

In 1962 it was demonstrated that the spectra of aqueous solution of α amino acids in which oxygen has been dissolved under high pressure showed marked increases in absorbance in the ultra-violet. This has been attributed

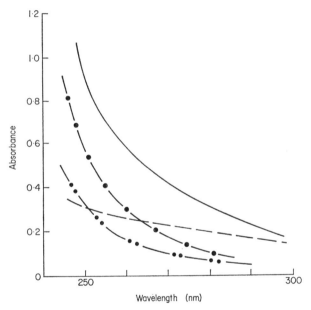

Fig. 3.2. Absorption spectra. Glycine in water ─────. Oxygen under 130 atmospheres in water ─··─··─. Oxygen and glycine in water under 130 atmospheres ─────. Difference between ─··─ and ─·─·─. After Fig. 1, ref. 3.

to a charge transfer spectrum (1). Although no formal attempt was made to measure enthalpies of dissociation, it was presumed that no bound complexes as such were formed, as the release of pressure above the oxygen caused an instantaneous bubbling off of the dissolved oxygen and the reversal of the spectrum to that of the amino acid. It was also shown that for solutions of aromatic hydrocarbons and oxygen which give rise to similar changes in spectra, Benesi-Hildebrand plots extrapolate back to the origin (2). This is contact charge transfer in which a donor molecule on passing close to an acceptor molecule momentarily donates charge to it. No bound complexes are formed and the distribution of the molecules in the solution is what one would expect from non-interacting molecules. The charge transfer from the amino acids must come from the lone-pair electrons on the amino group in the non-ionic form of the amino acid. Triethylamine and ethylamine show similar effects (3) but no effect is apparent with acetic acid used as the COOH analogue of the amino acids. This effect was not observed with proteins but this is due to the low solubility of proteins and in some cases the turbidity of the solutions.

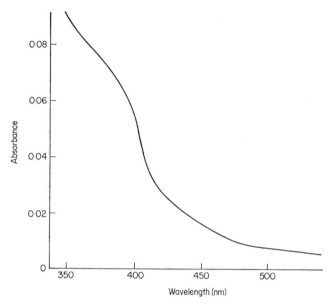

Fig. 3.3. Absorption spectrum of tryptophan and oxygen (130 atmospheres) in water minus absorption spectrum of tryptophan alone in water. Optical path length 5.6 cm. After Fig. 2, ref. 3.

Tryptophan shows a rather different spectral change from the other amino acids. The new absorption with oxygen lies to longer wavelength than with the aliphatic amino acids or amines and in addition there is a superimposed band on the new absorption believed to be the singlet-triplet transition of tryptophan at 360 nm. It may well be that tryptophan here functions as a π-donor instead of, or in addition to being an n-donor. Perturbation of the π-electron system would be required to make the singlet-triplet transition allowed but whether the proximity of the paramagnetic oxygen is sufficient to promote this transition without the π-electron being involved in charge transfer to the oxygen is a debatable point.

3.3. Chloranil

A series of studies have been carried out on the interaction of amino acids and proteins with the well-known organic electron acceptor chloranil (tetrachlorobenzoquinone). This molecule was chosen as a model for the biological quinones (4, 5, 6, 7, 8).

Spectral studies of solutions of mixtures of amino acids and chloranil in aqueous ethanol has revealed some interesting changes (4). The spectrum of

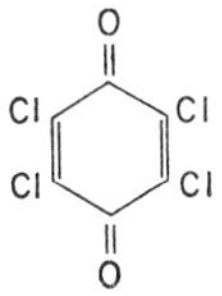

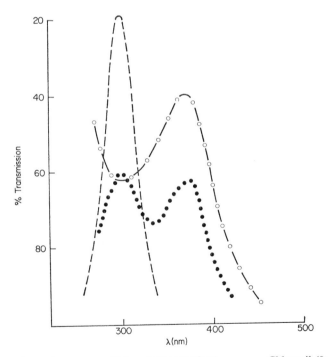

Fig. 3.4. Tetrachlorobenzoquinone (chloranil).

Fig. 3.5. Absorption spectra. Chloranil (8×10^{-5} M) — — —. Chloranil (8×10^{-5} M) and glycine (10^{-2} M) —○—○—. Chloranil (8×10^{-5} M) and phenylalanine (10^{-3} M) ●●●●●. All in 50% ethanol. After Fig. 1, ref. 4.

chloranil in aqueous ethanol has a strong absorbance at *ca.* 290 nm. On adding an amino acid, this peak decreases in intensity and shifts slightly to *ca.* 295 nm. A new peak occurs at longer wavelength between *ca.* 350 nm and 390 nm depending on pH. This new peak is slow to develop, some typical times being 9 hr at pH 9, 24 hr at pH 7 and 3 days at pH 4. The rate of growth of this new peak corresponds to the rate scheme:

$$D + A \underset{k_2}{\overset{k_1}{\rightleftharpoons}} (D : A)$$

Furthermore, at equilibrium the absorbance of the new peak obeys the Benesi-Hildebrand equation for 1 : 1 complexes. Molecular extinction coefficients predicted by the Benesi-Hildebrand equation substantially agree with values predicted by the kinetic method of following the rate of increase of absorbance (7). The effect of temperature on these spectra is to cause a reversal of the changes.

Solid complexes of amino acids and chloranil have been prepared by evaporating down old solutions in a rotary evaporator at low temperature

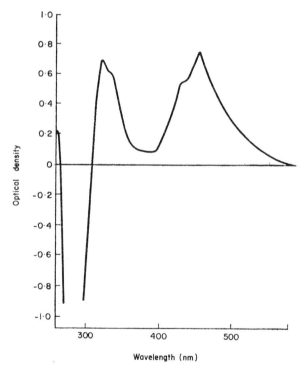

Fig. 3.6. Difference spectrum. Isoleucine and chloranil *vs.* chloranil in dimethyl sulphoxide. Reproduced with permission from Fig. 2, ref. 6.

TABLE 3.1

Complexes of amino acids with chloranil in aqueous solution

Amino acid	λ_{max} (nm)	pH	$K_c (M^{-1})$	Ref.
glycine	370	5.6	$298_{20°C}$	4
	360	8	$215_{20°C}$	4
	350	10	$127_{20°C}$	4
	340	11		4
leucine	360	7	$224_{20°C}$	4
alanine	355	8	$318_{20°C}$	4
tyrosine	360	7		4
	330	13.2		4
tryptophan	370	5		4
	360	7		4
	355	8	$176_{20°C}$	4
	340	10.4		4
	330	12	$ca.\ 0_{20°C}$	4
proline	330	1		7
	330	2		7
	330	4	$21_{22°C}$	7
	375	6		7
	375	7	$11_{22°C}$	7
	370	9	$154_{20°C}$	7
	forward reaction rate $1.81 . 10^{-2}\ M^{-1}\ sec^{-1}$†			
	back reaction rate $1.2 . 10^{-4}\ sec^{-1}$			
	340	10.2		
	355	12		7
	335	13		7

proline spectra obtained in Gallenkampf buffer, all others in Burroughs Wellcome buffer

† The forward and back reaction rates are those of the scheme on p. 58 and equal to $k_1(k/k')$ and $k_2(k'/k)$ respectively.

under reduced pressure (8). Dark brown solids are produced whose infra-red spectra in KBr discs have been studied. The spectra of these solids show the spectrum of the unionized amino acids $COOHCHRNH_2$ unlike that of amino acid in the free state which is $COO^-CHRNH_3^+$. The spectrum of the complexed chloranil is identical to that of chloranil complexed with hydroquinone, i.e. a classical charge transfer complex. Complexed chloranil shows a shift of the carbonyl band from 1690 to 1633 cm^{-1} (9). This is unequivocal evidence that the amino acids form 1 : 1 charge transfer complexes of quinhydrone type with chloranil. Moreover tryptophan and tyrosine normally regarded as good π-electron donors also behave as n-donors in complexing with chloranil.

The slow interaction of chloranil with the amino acids can be explained as follows. In solution, the amino acids exist almost entirely in ionized forms as shown schematically:

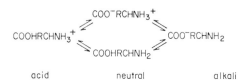

| acid | neutral | alkali |

Only the neutral $-NH_2$ can act as an electron donor as protonation locks up the lone-pair electrons. The ratio of unionized amino acid to zwitterion is very small indeed, about 1 part in 223,000 for glycine (10). The interaction scheme can be represented in a simplified form as follows:

$$AH^+ \underset{k'}{\overset{k}{\rightleftharpoons}} A, \qquad A+B \underset{k_2}{\overset{k_1}{\rightleftharpoons}} (A : B)$$

The concentration at the very beginning is given by

$$\frac{[A]}{[AH^+]} = \frac{k}{k'}, \text{ and at equilibrium by } \frac{[A_0]}{[AH^+ - A_0]} = \frac{k}{k'}$$

as $[AH^+] \gg [A]$ then the concentration of A is sensibly constant throughout the reaction and is very small. The rate of growth of the complex $(A:B$ is) proportional to the concentration of the reactants. Thus

$$\frac{d[(A : B)]}{dt} = k_1[A][B-(A : B)] - k_2[(A : B)];$$

at equilibrium

$$k_1[A_0][B-(A : B)_0] = k_2[(A : B)_0].$$

Eliminating k_2 between these two equations gives

$$\frac{d[(A : B)]}{dt} = k_1[A][B]\left(1 - \frac{[(A : B)]}{[(A : B)_0]}\right) \quad \text{as } [B] \gg [(A : B)]$$

or

$$\frac{d[(A : B)]}{dt} = k_1 \frac{k}{k'}[AH^+][B]\left(1 - \frac{[(A : B)]}{[(A : B)_0]}\right).$$

This equation has the identical form to that for the reaction $A + D \rightleftharpoons (A : D)$ (see Section 2.8) but will take very much longer to reach equilibrium as compared to straight forward charge transfer complexing, due to the very low value of $k/k'[AH^+]$.

The value of k/k' increases with increasing pH, due to the increasing amounts of COO^-RCHNH_2 formed. Hence the reaction takes less time to reach equilibrium at higher pH.

The spectra of amino acid chloranil mixtures in dimethyl sulphoxide are those of the chloranil anion (6) (see Fig. 3.6). Solutions are stable over many months whereas the sodium chloranil salt, whose spectrum is that of the chloranil anion, reverts to the spectrum of chloranil in three hours. The appearance of the stable anion of chloranil is interpreted as due to the formation of charge transfer complexes, with almost complete charge transfer in the ground state. It has not been proved possible to isolate the complexes as even prolonged heating under vacuum leaves a thick oily liquid.

TABLE 3.2

Complexes of proteins with chloranil

Protein	λ_{max} (nm)	pH
fibrinogin	ca. 360	5
	330	7
trypsin	ca. 340	5
	345	8
gamma globulin	ca. 340	7
	350	8
horse serum albumin	350	8

All compounds in 50% aqueous buffer and ethanol. Reproduced with permission from Table 3, ref. 4.

The behaviour of protein chloranil mixtures is similar to that of the amino acid chloranil mixtures (4), with the sole exception that the wavelengths of the complexed chloranil peak lie at shorter wavelength with the proteins than with the amino acids (see Fig. 3.7).

This new peak seen with both proteins and amino acids and whose position varies with pH is not the charge transfer band but an intrinsic band of chloranil shifted on complexing to that of the negative ion at 426 nm due to the contribution of ionic structures to the structure of the complexed ion (5). None of the criteria for the charge transfer band given in Section 1.17 apply in these cases. The fractional amount of shift of the complexed band has been suggested as a method of assessing the ratio of the coefficients in the Mulliken theory (5). There is a good correlation found between the amount the chloranil band is shifted to the red and the association constants as would be expected from the Mulliken theory.

Solid bovine serum albumin chloranil complexes have been prepared and have a higher conductivity than the protein alone (4). Similar results have been obtained by Snart et al. (11). They have interpreted such changes as arising from charge transfer from the valence band of the protein to the

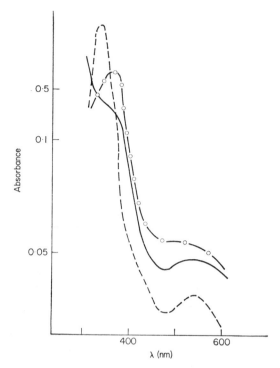

Fig. 3.7. Absorption spectra. Chloranil and trypsin —○—○—. Chloranil and horse serum albumin -----. Chloranil and γ globulin ———. All in 50% ethanol. After Fig. 6b, ref. 4.

chloranil, or other acceptors, which leaves behind a "hole" in the protein, thus increasing the current carrying capacity. Bovine serum albumin complexes with β-carotene, chlorophyll, β-naphthoquinone and various carcinogens with similar increase in conductivity (11).

3.4. Tetracyanoethylene

Bollard has studied the interactions of amino acid with another well-known acceptor tetracyanoethylene (TCNE) in anhydrous acetic acid (12). The addition of aliphatic amino acids or the aromatic phenylalanine to TCNE causes a long wavelength shift of the TCNE absorption bands. These changes occur slowly. By analogy with the interactions with chloranil which display very great similarities it is suggested that n-π charge transfer complexes are formed between the amino acids and TCNE. The use of the acid as solvent produces some problems as presumable the acid itself can act as an electron acceptor. Bollard also points out that TCNE discolours slowly in the acid.

The reversibility of these amino acid TCNE interactions have not been demonstrated nor have association constants been measured. The interaction of tryptophan with TCNE is rather different. Initially, on mixing the amino acid and the acceptor, a blue colour appears. In time this colour disappears as the spectrum changes to that of the other amino acid mixtures. It is suggested that the initial colour change is due to the formation of a π-complex but this gradually changes over to an n-complex. This result is consistent with the results obtained with chloranil. If both the indole ring and the unionized amino group of the amino acid can complex with TCNE then the initial reaction must be with the indole ring as the amount of unionized amino acid present is very small. With time, the interaction of the lone-pair electrons of the unionized amino acid will cause a shift of equilibrium to the unionized form. If then the n-π interaction is stronger than the π-π interaction, the π-π complex will be displaced in time by the n-π complex. The non-interaction of phenylalanine as a π donor is simply due to the phenyl system of the amino acid being a much weaker donor than the indole ring of tryptophan.

3.5. Haematoporphyrin

Another series of interactions is with haematoporphyrin. The porphyrins as a group are discussed in Chapter 6. Haematoporphyrin exhibits different

TABLE 3.3

Complexes of amino acids with haematoporphyrin[a]

Amino Acid	$\Delta H°$ (kcal/mole)	K_c (M^{-1})	T (°C)
tryptophan	− 5·2	7·18	21
6-aminocaproic acid	− 5·5	1·13	21
arginine	− 5·0	12·4	24

In pH9 buffer.
[a] Ref. 13.

visible spectra at different pH. In unbuffered 50% aqueous ethanol, the effect of adding amino acids is to change the spectrum of the porphyrin but the same identical changes can be produced by using acid or alkali to the same pH. Interestingly enough, the change in spectrum caused by adding an amino acid obeys the Benesi-Hildebrand equation and apparent association constants can be obtained for the amino acids (13, 14). These apparent association constants closely parallel the electron donating ability of the amino acids and aliphatic amines, although they measure the dissociation of these molecules. This demonstrates the close relationship between the basicity of a substance and its electron donating ability.

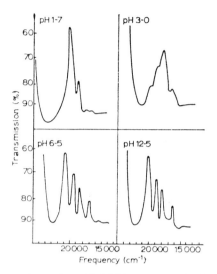

Fig. 3.8. Absorption spectrum of *ca.* 10^{-5} M haematoporphyrin at different pH. Reproduced with permission from Fig. 7, ref. 13.

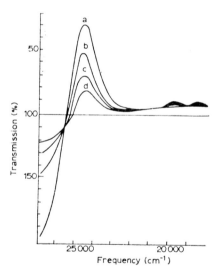

Fig. 3.9. Difference spectra. (a) 65.7 mM tryptophan in 20.9 μM haematoporphyrin; (b) 32.9 mM tryptophan in 20.9 μM haematoporphyrin; (c) 19.7 mM tryptophan in 20.9 μM haematoporphyrin; (d) 13.1 mM tryptophan in 20.9 μM haematoporphyrin in pH 9 buffer. Reproduced with permission from Fig. 9, ref. 13.

In buffered solutions at pH 9, spectral changes are observed in certain amino acid haematoporphyrin solutions which are temperature reversible and for which association constants and enthalpies of dissociation have been evaluated. The following amino acids have been found to form complexes, tryptophan, 6-amino caproic acid, arginine, histidine and phenylalanine with $\Delta H°$ of *ca.* 5 kcal/mole. None of the aliphatic α amino acids react however with haematoporphyrin.

It is suggested that as indole does not react with haematoporphyrin under these conditions but amino acids do, all with the same enthalpy of dissociation, then tryptophan must here be acting as an *n*-donor rather than a π-donor. The non-interaction of the α amino acids is unexplained.

3.6. Vitamins B_{12} and B_{12b}

A molecule of great medical interest is vitamin B_{12} (cyanocobalamin) and one of its derivatives, vitamin B_{12b} (hydroxy- or aquocobalamin). This vitamin has proved very efficacious in the treatment of megablastic anaemia. Clinical tests have shown that pure crystalline B_{12} is not as effective medically as less pure forms of the vitamin obtained during manufacture (15). These impure forms are believed to contain either amino acids or other low molecular weight peptides or alternatively proteins (the extrinsic factor). Studies by Pullman suggest that vitamin B_{12} should be a good electron acceptor (16). Experimental studies however indicate that vitamin B_{12} is an electron donor.

TABLE 3.4

Interaction of amino acids and peptides with vitamin B_{12b}[a]

Compound (10^{-1} M)	Change in absorbance between 370 nm and 348 nm
glycine	0·594
alanine	0·468
serine	0·456
proline	0·144
glycylglycine	0·566
glycylalanine	0·599
glycylserine	0·546
glycylproline	0·493
alanylglycine	0·343
alanylalanine	0·172
alanylserine	0·113
serylglycine	0·283
prolylglycine	No measurable spectrum

In aqueous buffer.
[a] Ref. 17.

This is discussed in Section 6.9.1. Studies of the interaction of vitamin B_{12b} with amino acids and peptides have revealed a weak interaction between these molecules.

Ultra-violet and visible spectrophotometric studies and kinetic studies show that the amino acids form 1 : 1 complexes with the vitamin (14, 17). Infra-red studies of solid glycine: B_{12b} complex show that the amino acid is

TABLE 3.5

Complexes of glycine with vitamin B_{12b}[a]

Temp.	pH	Forward reaction[b] rate $M^{-1}sec^{-1}$	Back reaction[b] rate sec^{-1}	$K_c (M^{-1})$
25°C	7	.04	$6.8 . 10^{-4}$	170
25°C	5	$3.8 . 10^{-4}$	$6.5 . 10^{-5}$	5.7

In aqueous buffer.
[a] Ref. 17.
[b] The forward and back reaction rates are those of the scheme on p. 58.

in the complex in its non-zwitterionic form, a situation analogous to the amino acid chloranil complexes (8). Again it is suggested that complexing takes place via n-electron donation from the nitrogen in the amino group of the amino acid. The relative complexing ability of the amino acids is glycine > alanine > serine > proline (17). These differences possibly arise from steric hindrance. A similar study of dipeptides has shown that their complexing ability is determined primarily by the N-terminal group although the C-terminal group has some slight effect on complexing in the same order as for the amino acids (17). Several large molecular weight peptides, i.e. polyglycine, polyproline and bovine serum albumin also complex with vitamin B_{12b} (17). It cannot therefore be assumed that the effectiveness of impure B_{12b} preparations in clinical studies is necessarily due to the presence of amino acids rather than proteins. Examples of spectra of the vitamin are given in Chapter 6.

3.7. Riboflavin

The flavins are the subject of Chapter 7 but their interaction with the aliphatic amino acids are discussed here.

The addition of the amino acids to riboflavin in a fully oxidized form causes the spectrum to change to that of a partially reduced form (18). Conversely, the presence of these amino acids in riboflavin solutions, protect

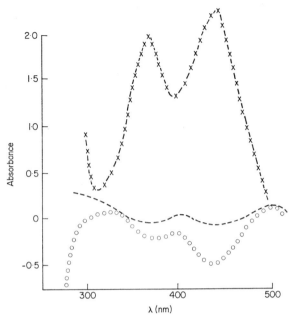

Fig. 3.10. Absorption spectrum of RFN in aqueous buffer — × — × —. Difference spectra. RFN plus phenylalanine *vs.* RFN ○ ○ ○ ○ ○. RFN plus leucine *vs.* RFN -----. After Fig. 1, ref. 18.

the riboflavin against further reduction. Beinert (19) has shown that the reduction of riboflavin is a two electron process, i.e.

oxidized riboflavin ⇌ semiquinone riboflavin ⇌ reduced riboflavin.

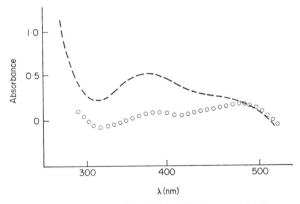

Fig. 3.11. Absorption spectrum of old RFN in pH 9 aqueous buffer -----. Difference spectrum of RFN plus glycine *vs.* RFN ○ ○ ○ ○ ○. After Fig. 2, ref. 18.

The action of the amino acids is to stabilize the semiquinone by one electron donation to the oxidized flavin on forming a complex. This opinion has been challenged by Kosower (20) who interprets the results as due to hydrophobic bonding of the amino acids to the flavin, the spectral changes being simply perturbation effects. As the spectrum gives the same isosbestic points as those reported by Beinert (19) for riboflavin during the process of reduction, the first theory seems the more probable especially as the amino acids are known electron donors and the flavins good acceptors.

3.8. Iodine

Solutions of amino acids with iodine in water exhibit spectral changes which proceed in two stages (6). Initially on mixing, the spectrum of iodine

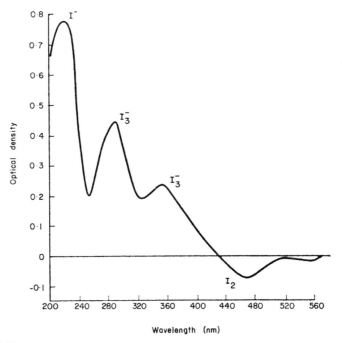

Fig. 3.12. Difference spectrum. Iodine *vs.* isoleucine and iodine. Immediately on mixing. Reproduced with permission from Fig. 11, ref. 6.

I_2 decreases in intensity and there is an accompanying increase in the absorption bands associated with the triiodide ion I_3^-. The second slower stage of change shows the gradual loss of both I_2 and I_3^- absorption and the appearance of the absorption spectrum of the iodide ion I^-. Similar results have been obtained for the interaction of other n-electron donors with iodine in water (21). The first stage is the formation of a charge transfer complex,

$$\text{amino acid} + 2I_2 \rightleftharpoons (\text{amino acid} : I)^+ + I_3^-.$$

The second stage is a later chemical reaction between the amino acid and iodine in the complex leading to some form of amino acid iodide. The proteins, γ globulin and bovine serum albumin behave in a similar manner.

3.9. Intramolecular Complexes

Complexes between electron donors and acceptors suitable for incorporation into artificial peptides have been studied by Carrion et al. Only π-π charge transfer complexes have been studied, the authors pointing out that while n-donation may occur from free amino groups, the lone-pair electrons in the peptide bond do not take part in charge transfer nor do the lone-pair electrons on carboxylated or acetylated amino groups (22, 23).

This view is confirmed by Buvet (24) who has shown that only the very strong electron acceptor bromine will complex with N-methyl acetamide (a model for the peptide bond) but no other weaker acceptors. Carrion and his coworkers have demonstrated that coloured complexes are formed between various phthalimides and substituted phenylalanines or more conventional donors such as indole and dimethylaniline (23). Coloured compounds are formed when phthalyltryptophan is incorporated into the same molecules as phenylalanine, the colour being ascribed to an intra-molecular charge transfer complex (see Tables 3.6 and 3.7, and Figs 3.13 and 3.14).

Moser (25) has studied the circular dichroism and optical rotatory dispersion associated with the charge transfer bands of the inter and intramolecular complexes studied by Carrion et al. (22, 23). Optically active molecules have circular dichroic bands associated with the charge transfer bands. Using the area under the bands as a measure of complexing, association constants have been evaluated which agree with those obtained from the Benesi-Hildebrand equation applied to absorbance changes.

3.10. Cysteine and Glutathione

Mixtures of two sulphur-containing amino acids, cysteine and glutathione with chloranil in aqueous dioxane, exhibit ESR signals (26). It is suggested that there are two consecutive electron transfer steps involving the sulphur of the amino acids resulting in the reduction of chloranil via the semiquinone to the hydroquinone.

TABLE 3.6

Charge transfer complexes of 4-nitrophthaloylglycine ethyl ester

Acceptor:

No.	Donor	Solvent	Colour	λ_{max} (nm)	ε_{DA}	K_c (M^{-1})
1		Acetone	yellow	347	643	1.1
		HCCl$_3$	yellow	383		
2		HCCl$_3$	yellow	360	220	1.1
		EtOH	yellow	370	500	2.3
3		EtOH	red/brown	450	555	0.57

No.	Structure	Solvent	Colour			
4	CH_3, CH_3-N-〈 〉-CH_3	EtOH	red/brown	470	710	0.56
5	CH_3, CH_3-N-〈 〉-CH_2CH($COOC_2H_5$)($NH \cdot COCH_3$)	EtOH	reddish	460	664	0.54
6	indole (NH)	Acetone	yellow	360	1280	0.25
7	CH_2CH(COOH)($NH \cdot CO \cdot OCH_2$-〈 〉) on indole	EtOH	yellow	355	712	1.56

Properties of some intermolecular charge transfer complexes (25°)

Reproduced with permission from Table 1, ref. 22.

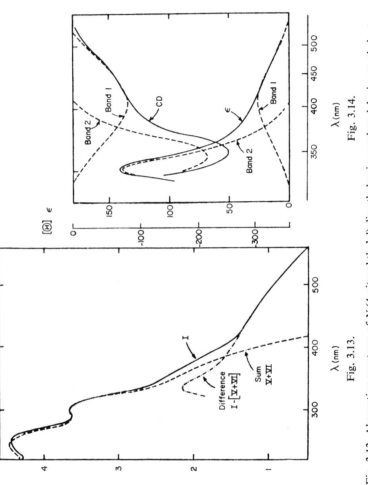

Fig. 3.13.

Fig. 3.14.

Fig. 3.13. Absorption spectrum of N-(4-nitrophthalyl) dimethylamino-L-phenylalanine methyl ester (I) ——. Sum of the spectra of N-(nitrophthalyl)-glycine ethyl ester (VI) and dimethyl amino toludine (V) ———. Difference between spectrum and sum of (V) and (VI) —·—·— in ethanol at 25°.

Fig. 3.14. Extinction and circular dichroism of the intramolecular charge transfer band of (I). Reproduced with permission from Figs 1 and 2, ref. 13.

TABLE 3.7

Charge transfer complexes of phthalimides

Acceptor	Donor	λ_{max} (nm)	$\Delta G°$	$\Delta H°$ kcal/mole	$T\Delta S°$	K_c (M^{-1})	
4-nitrophthal-glycine-ethylester	dimethyltoluidine	475	0·3	− 1·7	− 2·0	0·6	A
4-nitrophthal-glycine-ethylester	indole	sh	0	− 1·8	− 2·6	0·97	B
tetrachlorphthalyl-glycine-methylester	hexamethylbenzene	360	0·6	− 2·1	− 2·7	0·4	B
tetrachlorphthalyl-glycine-methylester	dimethyltoluidine	470	0·5			0·4	B
tetrachlorphthalyl-glycine-methylester	dimethyltoluidine	455	0·2	− 0·6	− 0·8	0·9	C
tetrachlorphthalyl-glycine-methylester	indole	sh	− 0·1	− 2	− 1·9	1·2	B
tetrachlorphthalyl-glycine-methylester	indole	sh	0·1			0·8	C
4-nitrophthalylethylester	benzyloxylcarbonyl pentamethyl-phenyllalanine	370					A and B
4-nitrophthalylethylester	acetyl dimethylamino phenylalanine	458				0·7	B
4-nitrophthalylethylester	acetyl tryptophan	sh					D
4-nitrophthalimide	acetyl dimethyl aminophenylalanine	450					A
4-nitrophthalimide	benzyloxycarbonyl pentamethyl-phenyllalanine	368				2·8	A
phthalylglycine-methylester	dimethyltoluidine	sh					C

A = ethanol, B = chloroform, C = ethylacetate, D = dimethylformamide, sh = shoulder. Adapted from Tables 3 and 4, ref. 23.

3.11. D-Amino Acid Oxidase

A series of investigations have been carried out by Yagi and his collaborators on the interaction of the flavoprotein D-amino acid oxidase with amino acids.

A mixed solution of the protein with D-lysine under anaerobic conditions yields a purple complex with an absorption maximum at 550 nm. On adding

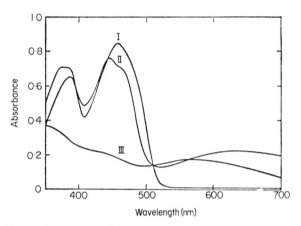

Fig. 3.15. Changes in spectrum of D-amino acid oxidase by the reaction with D-lysine. I: Oxidized state of D-amino acid oxidase, 7.44×10^{-5} M in respect to FAD; II: I was mixed with 1.5×10^{-4} M D-lysine under aerobic conditions, green complex; III: I was mixed with 1.5×10^{-4} M D-lysine under anaerobic conditions, purple complex. Reproduced with permission from Fig. 1, ref. 27.

oxygen, the purple complex changes colour to green with an absorption maximum at 630 nm (27, 32). Neither of these complexes give rise to ESR signals. They do however exhibit optical rotatory dispersion (29, 32).

Similar purple complexes have been isolated from mixtures of D-amino acid oxidase and D-proline and D-alanine (28–31). These complexes all have 1 : 1 stoichiometry. The absorption maximum of the alanine complex lies to somewhat shorter wavelength than that of the proline complex which in turn lies to shorter wavelength to that of the lysine complex. The ionization potential of the amino acid proline is about 0.27 eV smaller than that of alanine, as given by the difference between the maxima of the purple bands converted to energy. Slifkin and Allison have shown that the difference between the ionization potentials of these compounds is 0.27 eV as determined from contact charge spectra in the presence of oxygen (33).

These purple complexes are diamagnetic, i.e. they do not give rise to an ESR signal but on standing in the dark for several days an ESR signal appears, this is accompanied by a change in the absorption spectrum, the peak at

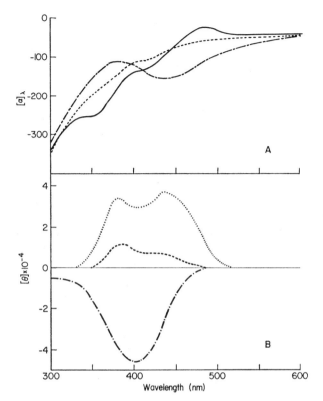

Fig. 3.16. Optical activity of D-amino-acid oxidase and its complexes. (A) Optical rotatory dispersion spectra. (B) Circular dichroism spectra. D-Amino-acid oxidase – – – – – The green complex produced through aerobic reaction with D-lysine ———. The purple complex produced through anaerobic reaction with D-lysine —· —· —. Reproduced with permission from Fig. 8, ref. 29.

550 nm being replaced by a peak at 492 nm believed to be due to the conversion of the purple complex to the semiquinoid enzyme (29). There is also a marked difference in spectrum between the purple complex and that of the semiquinone obtained by partial reduction of the enzyme. There is a clear difference in the ORD spectra of the purple complex and the semiquinoid enzyme.

The above results are compatible with the formation of n-π charge transfer complexes between the lone-pair electrons on the nitrogen in the amino acid and the isoalloxazine ring of the enzyme. The suggested structure of the complexes is $FAD^{-}.NH_{2}^{+}RCHCOO^{-}$. This is a similar structure to the amino acid chloranil complex which has been confirmed by infra-red spectroscopy (8) except that in solution, the purple complex is hydrolyzed. Ionization of the complex is considerably prevented by the hydrophobic environment of the enzyme.

Kinetic studies using the stopped-flow technique shows that the reaction scheme producing the purple complex is

$$E_{ox} + A \rightleftharpoons E_{ox} \ldots A \rightleftharpoons \text{purple complex} \rightleftharpoons E_{red} \ldots P \rightleftharpoons E_{red} + P$$

where E is the enzyme and A the amino acid (31), or in terms of the structures

$$FAD^- . NH_2^+ RCHCOO^- \dagger \rightleftharpoons FAD^- - NH_2 + COO^- RC :^- +$$
$$+ H^+ \rightleftharpoons FADH_2 + HNCRCOO \qquad \text{ref. 28.}$$

Yagi et al. have described the purple complex as an inner complex, i.e. one which is primarily dative in the ground state and the green complex an outer complex, one where the ground state possess little dative character.

Under aerobic conditions, the addition of the amino acids to the enzyme causes chemical changes to occur without the appearance of the purple intermediate which can be represented thus:

$$E_{ox} + S \rightleftharpoons E_{ox} \ldots S \rightleftharpoons E_{red} \ldots P \overset{O_2}{\rightleftharpoons} E_{ox} . P \rightleftharpoons E_{ox} + P$$

or possibly

$$E_{ox} + S \rightleftharpoons E_{ox} \ldots S \rightleftharpoons E_{red} \ldots P \rightleftharpoons E_{red} + P \overset{O_2}{\rightleftharpoons} E_{ox} + P \rightleftharpoons E_{ox} \ldots P \qquad \text{ref. 34}$$

3.12. N-Methylphenazonium Sulphate

Recently it has been demonstrated that the addition of various molecules of biological interest containing lone-pair electrons, including glycine to solutions of N-methylphenazonium sulphate gives rise to free radical signals (35). The solutions also change colour. These effects are interpreted as arising from charge transfer interactions involving the lone-pair electrons on the amino and possibly the carboxyl groups of the amino acid.

3.13. Conclusions

Amino acids and proteins can form charge transfer complexes via the lone-pair electrons on the terminal nitrogen in the unionized form of the amino acid or peptide. It is therefore important in considering the role of charge transfer forces in biology to bear in mind that n-electron donation cannot be dismissed as a mode of interaction.

† Yagi in a private communication now believes the structure to be
$$FAD_2^- . NH_2^+ RCCOO^-.$$

REFERENCES

1. Slifkin, M. A. (1962). *Nature*, **193**, 464.
2. Evans, D. F. (1962). *J. chem. Soc.* 1987.
3. Slifkin, M. A. (1962). Ph.D. Thesis, Manchester University.
4. Birks, J. B. and Slifkin, M. A. (1963). *Nature*, **197**, 42.
5. Slifkin, M. A. (1963). *Nature*, **198**, 1301.
6. Slifkin, M. A. (1964). *Spectrochim. Acta*, **20**, 1543.
7. Slifkin, M. A. and Heathcote, J. G. (1967). *Spectrochim. Acta*, **23A**, 2893.
8. Slifkin, M. A. and Walmsley, R. H. (1969). *Experientia*, **25**, 930.
9. Slifkin, M. A. and Walmsley, R. H. (1970). *Spectrochim. Acta*, **26A**, 1237.
10. Mahler, H. R. and Cordes, E. H. (1964). "Biological Chemistry," p. 194. Harper, London.
11. Davies, K. M. C., Eley, D. D. and Snart, R. S. (1960). *Nature*, **188**, 724.
    Eley, D. D. and Snart, R. S. (1965). *Biochim. biophys. Acta*, **102**, 379.
    Snart, R. S. (1968). *Biopolymers*, **6**, 73.
12. Bollard, J. (1969). *J. chim. Phys.* **69**, 221.
13. Heathcote, J. G., Hill, G. J., Rothwell, P. and Slifkin, M. A. (1968). *Biochim. biophys. Acta*, **153**, 13.
14. Slifkin, M. A. and Heathcote, J. G. (1968). "Molecular Associations in Biology" (Ed. B. Pullman). Academic Press, New York.
15. Mooney, F. S. and Heathcote, J. G. (1967). *Lancet*, **11**, 397.
16. Veillard, A. and Pullman, B. (1965). *J. Theoret. Biol.* **8**, 307.
17. Heathcote, J. G. Moxon, G. H. and Slifkin, M. A. *Spectrochim Acta* (in press).
18. Slifkin, M. A. (1963). *Nature*, **197**, 275.
19. Beinert, H. (1956). *J. Am. chem. Soc.* **47**, 114.
20. Kosower, E. M. (1965). "Flavins and Flavoproteins" (Ed. E. C. Slater). Elsevier, Amsterdam.
21. Slifkin, M. A. (1965). *Spectrochim. Acta*, **21**, 1391.
22. Carrion, J. P., Donzel, B., Deranleau, D. A., Esko, K., Moser, P. and Schwyzer, R. (1967). "Peptides." North Holland, Amsterdam.
23. Carrion, J. P., Deranleau, D. A., Donzel, B., Esko, K., Moser, P. and Schwyzer, R. (1968). *Helv. Chim. Acta*, **51**, 459.
24. Buvet, R. (1969). Abstr. 3rd Int. Biophys. Congress. Cambridge, USA.
25. Moser, P. (1968). *Helv. Chim. Acta*, **51**, 1831.
26. Gause, E. M., Montalvo, D. A. and Rowlands, J. R. (1967). *Biochim. biophys. Acta*, **141**, 217.
27. Yagi, K., Kotaki, A., Naoi, M. and Okamura, K. (1966). *J. Biochem.* **60**, 236.
28. Yagi, K. and Nishikimi, M. (1969). *Biochem. biophys. Res. Commun.* **34**, 549.
29. Yagi K., Okamura, K., Naoi, M., Sugiura, N. and Kotaki, A. (1967). *Biochim. biophys. Acta*, **146**, 77.
30. Yagi, K. and Ozawa, T. (1964). *Biochim. biophys. Acta*, **81**, 29.
31. Yagi, K., Nishikimi, M., Ohishi, N. and Hiromi, K. (1969). *J. Biochem.* **65**, 663.
32. Yagi, K., Okamura, K., Naoi, M., Takai, A. and Kotaki, A. (1969). *J. Biochem.* **66**, 581.
33. Slifkin, M. A. and Allison, A. C. (1967). *Nature*, **215**, 949.
34. Yagi, K., Ozawa, T. and Naoi, M. (1969). *Biochim. biophys. Acta*, **185**, 31.
35. Kimura, J. E. and Szent-Györgyi, A. (1969). *Biochem.* **62**, 286.

CHAPTER 4

Purines and Pyrimidines

4.1. Introduction

A large portion of the work on biological charge transfer has been carried out on purines and pyrimidines. This is undoubtedly due to the interest in the last few years in the role of nucleic acids in genetics. Five purines and pyrimidines, *viz.* guanine, adenine, uracil, thymine and cytosine constitute the base pairs of DNA and RNA. However other purines and pyrimidines do have biological importance. Caffeine for example is a well-known stimulant.

4.2. Aromatic Hydrocarbons

The earliest work in this area was carried out to determine whether the purine or pyrimidines complex with polycyclic aromatic hydrocarbons, many of which are potent carcinogens, at least to small rodents if not to human beings. As these hydrocarbons form 1 : 1 charge transfer complexes with conventional organic acceptors, it was natural that workers in this field should consider the possibility of charge transfer complexing with biological molecules.

As early as 1938, well before the idea of charge transfer complexing had been mooted, Brock and coworkers (1) showed that caffeine had a marked solubilizing effect on various polycyclic compounds in water. Weil-Malherbe (2) measured the solubilizing power of different purines on a number of hydrocarbons and found that the solubilizing power of the purines increased with increasing *N*-methylation, so that 1, 3, 7, 9-tetramethyluric acid was the best solubilizing agent of the purines studied. Weil-Malherbe was also able to isolate crystalline compounds containing molecules of purines and hydrocarbon in well-defined molecular ratios. The tetramethyluric acid pyrene mixed crystal contained the molecules in 1 : 1 stoichiometry. The tetramethyluric acid 3,4-benzopyrene mixed crystal has a 2 : 1 stoichiometry. These results of Weil-Malherbe were confirmed by the X-ray crystallographic studies of Liquori *et al.* (3), who showed not only that the stoichiometry was 1 : 1 but that the two molecules were arranged in the crystal in a sandwich configuration, similar to those found for crystals of well-known charge transfer complexes. The intermolecular separation is 3·4 Å (see Fig. 4.2).

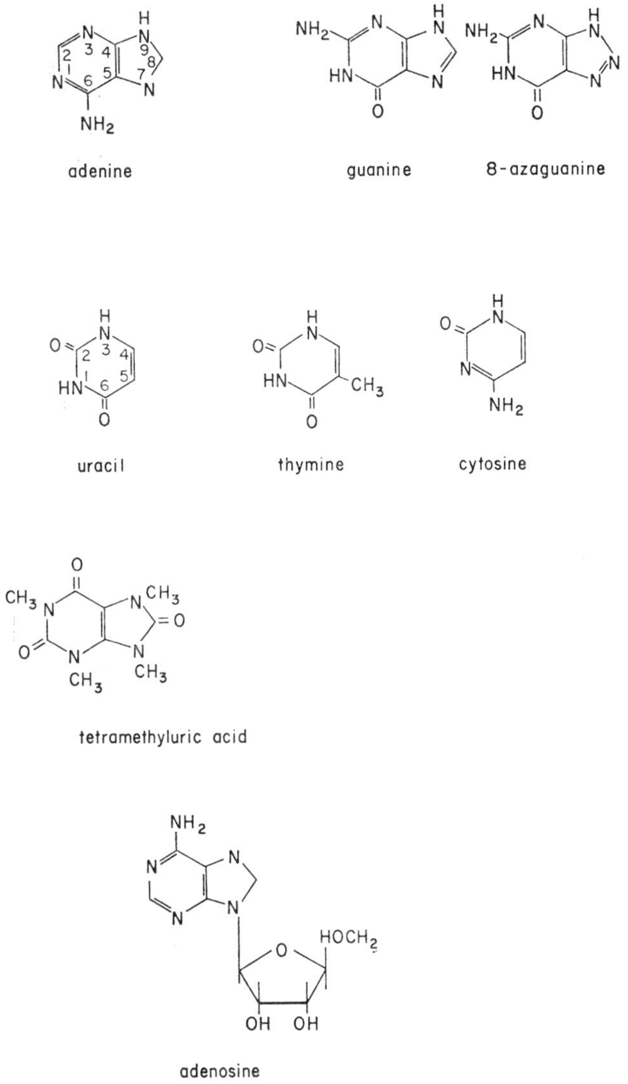

adenine   guanine   8-azaguanine

uracil   thymine   cytosine

tetramethyluric acid

adenosine

Fig. 4.1. Some purines and pyrimidines.

A series of studies of hydrocarbon purine complexes have been carried out by workers at the Chester Beatty Hospital. Booth and Boyland (4) and with others (5) have examined the solubilities of various aromatic compounds in aqueous solutions. In common with earlier workers, they observed marked increases in solubilities in purine solutions as compared to water. They were also able to isolate from these solutions solid molecular complexes or compounds containing the components in simple stoichiometry. The absorption spectra of the hydrocarbons in purine solutions are similar in form to those in 50% aqueous ethanol but are slightly shifted to longer wavelength and

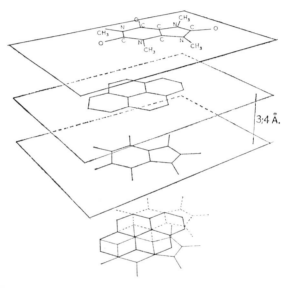

Fig. 4.2. Crystal structure of the tetramethyluric acid pyrene complex. Schematic drawing of the perpendicular separation of the molecules in one stack through the unit cell. Reproduced with permission from Fig. 2, ref. 3.

have a lower absorbance. The fluorescence of the hydrocarbons is quenched in the presence of acidified caffeine solution and in tetramethyluric acid at all pH's, a similar result to that found by Weil-Malherbe (3, 6) (see Fig. 4.4). It was concluded that the interaction was of an indefinite nature possibly involving hydrogen-bonding. The infra-red spectra of these purine hydrocarbon complexes were studied by Booth, Boyland, and Orr (7) to see whether the interaction could be characterized. The infra-red bands associated with the carbonyl $-CO$ groups in the purines were invariably red shifted in the complex. Bands associated with $-C=N$ bonds contiguous to the carbonyl groups were invariably blue shifted in the complex. Strong aromatic bands of

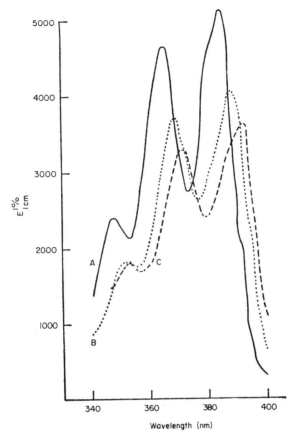

Fig. 4.3. Ultra-violet absorption data. (A) 3 : 4-Benzopyrene in 50% EtOH. (B) 3 : 4-Benzopyrene in 50% EtOH saturated with caffeine. (C) 3 : 4-Benzopyrene in 50% EtOH saturated with tetramethyluric acid. Reproduced with permission from Fig. 2, ref. 7.

the hydrocarbons arising from C – H deformation modes were found to be blue shifted also (see Fig. 4.5). These results taken together with the foregoing were interpreted as due to an attraction between the two components arising from the polarization of the hydrocarbon by the polar purine. The carbonyl band shift is not inconsistent with the formation of a charge transfer complex between the two components although there does not appear to be a shift of the C = C bands which one would expect if there were any decrease in strength of these bonds due to charge transfer into an anti-bonding orbital.

Further work on these systems has been carried on by Boyland and Green (8). They have examined the solubilizing power of different purines to

TABLE 4.1

The solubility of polycyclic hydrocarbons in aqueous purine and pyrimidine solutions

Purine or pyrimidine	Purine or pyrimidine concentration (μM)	Hydro-carbon solubilized (μM)	M.R. purine — hydro-carbon	M.R. (caffeine) $\times$ 100 — M.R. (purine)
Benzo(a)pyrene				
6-Dimethylaminopurine	12 250	0.374	32 800	10.1
6-Methylaminopurine	6 710	0.078	86 000	5.4
	5 000	0.047	106 400	5.0
Guanine (in N H$_2$SO$_4$)	10 000	0.115	87 000	4.5
Guanosine	3 400	0.018	189 000	3.5
Hypoxanthine	5 000	0.029	172 500	3.2
Adenine	6 000	0.021	286 000	1.8
Inosine	6 000	0.020	300 000	1.7
Adenosine	10 000	0.031	323 000	1.3
Orotic Acid (pH 11·8 (NaOH))	5 000	0.007	714 000	0.75
Thymidine	30 000	0.080	375 000	0.54
Cytidine	12 070	0.013	928 000	0.34
Uracil	30 000	0.007	4 287 000	0.05
Tryptophan	50 000	< 0.48	—	< 1.0
Urea	6 M	0.388	15 466 000	—
Pyrene				
Guanine (in N HCl)	10 000	4.43	2 260	11.7
Adenine	6 000	0.65	9 230	3.8
Tryptophan (pH 6·5)	50 000	0.8	62 500	0.13
DPN	20 000	5.16	3 880	3.97
3-*Fluoro*-10-*methyl*-1,2-*benzanthracene* (water solubility 0.019 μM)				
Caffeine	5 000	0.19	26,300	—
Caffeine	60 000	18.33	3 270	—
4-*Fluoro*-10-*methyl*-1,2-*benzanthracene* (water solubility 0.019 μM)				
Caffeine	5 000	0.74	6 710	—
Caffeine	60 000	41.7	1 440	—

Reproduced with permission from Table 1, ref. 8.

aromatic hydrocarbons at different concentrations. Their results were similar to Weil-Malherbe's (6) with the exception that whereas Weil-Malherbe had found a change of slope in the solubilization curve, i.e. plot of hydrocarbon solubilized *vs.* purine concentration and interpreted this as a change from 1 : 1 complexing at higher purine concentration, Boyland and Green found no such break in the curve. However fluorescence quenching studies showed

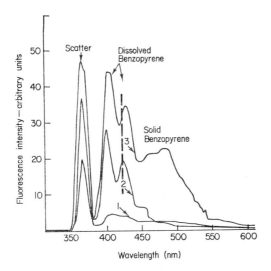

Fig. 4.4. Fluorescence spectra of 0.002 M aqueous caffeine solution shaken with solid 3,4-benzpyrene. Activation wavelength 365 nm. Curve 1, shaken 30 sec; Curve 3, shaken 30 min; Curve 2 3,4-benzpyrene in ethanol for comparison. Reproduced with permission from Fig. 1, ref. 8.

TABLE 4.2

Correlation of the solubilizing power of tetramethyluric acid for various hydrocarbons and their ionization potentials and electron affinities

Decreasing order of solubilization of hydrocarbon[a]	Ionization potential[b] (eV)	Electron affinity[c] (eV)
phenanthrene	8.09	0.69
pyrene	7.55	1.23
3,4-benzopyrene	7.19	1.59
1,2-benzanthracene	7.45	1.33
anthracene	7.37	1.41
chrysene	7.80	0.98
9,10-dimethyl 1,2-benzanthracene	7.43	
coronene	7.44	1.34
1,2,7,8-dibenzanthracene	7.68	1.10

[a] From ref. 2, with permission.
[b] From Birks, J. B. and Slifkin, M. A. (1961). *Nature*, **191**, 761.
[c] From ref. 51a, with permission.

that at low tetramethyluric acid concentrations, 1 : 1 complexes were formed with the hydrocarbons pyrene and 3,4-benzopyrene. At higher concentrations 2 : 1 complexes were formed. These studies were taken to reinforce the idea that complexing arises mainly from polarization forces. A point of some interest is that no solubilizing effect was observed with pyrimidines.

Van Duuren (9) has carried out fluorescence studied on similar systems. The fluorescence spectra of various hydrocarbons complexed with 1,3,7,9-tetramethyluric acid was observed both in solution and in KBr discs. The infra-red spectra of these discs were also obtained. In solution there was no difference between the fluorescence of the mixtures and the sum of the fluorescence spectra of the individuals, thus showing that any complex is dissociated completely in solution. Solid state spectra are complicated to some extent by some of the hydrocarbons, pyrene for example, exhibiting excimer emission rather than normal fluorescence. An excimer is a short-lived dimer formed between an excited molecule and a ground state molecule at an optimum distance of about 4 Å with the two molecules coplanar and which is recognized by a characteristic emission on dissociating. In the crystalline state it happens that pyrene molecules are held at the optimum separation and orientation for excimer formation, so that the emission of the crystal is that of the excimer rather than the normal fluorescence as observed in weak solution. In the tetramethyluric acid pyrene complex, the emission spectrum is blue-shifted, i.e. from the excimer emission towards the normal fluorescence. There is no evidence of any charge transfer emission bands. It is suggested that the change in emission is due to a short range interaction between the pyrene and purine of the polarization bonding type. Obviously the interpolation of the purine between pyrene molecules in the solid state by what ever mechanism will prevent excimer emission. Other hydrocarbons showing red-shifts in emission when complexing with the purine are coronene and 3,4-benzopyrene. An argument advanced for these complexes not being of the charge transfer type, is that there is poor correlation between the solubilizing power of tetramethyluric acid towards the hydrocarbons and their electron affinities (10) which would be expected if the major stabilizing force was charge transfer and if the purine was the donor (but see 4.16).

The writer has prepared some aromatic hydrocarbon purine complexes by evaporation of equimolar solutions of purine and hydrocarbon in aqueous acetone. The resulting residues show the characteristic red shift of the hydrocarbon absorption spectra. No effect is observed at all on the triplet-triplet absorption spectra or triplet lifetimes of the hydrocarbons. This behaviour is quite unlike that of complexes of the hydrocarbons with chloranil which display a marked quenching of the triplet state lifetime and a blue shift of the triplet-triplet spectrum (11).

Various theoretical studies have been carried out by Pullman and his coworkers (12). They conclude that stability of the complexes is mainly due to dispersion forces although some correlation is observed between solubilization and the theoretical ionization potentials of the purines (see Fig. 4.16). The most thorough values for the solubilization of hydrocarbons by purines are those of Mold (13) and his fellow workers using a selective separation of the polycyclic aromatic hydrocarbons by counter current distribution.

4.3. Chloranil

There have been several studies of the interactions with well-known organic electron acceptors. Beukers and Szent-Györgyi (14) looked at the ultra-violet and visible spectra of mixtures of chloranil with thymine or adenine, and 1,3,5-trinitrobenzene with adenine, all in dimethyl sulphoxide. Although purple coloured solutions were obtained with chloranil, these probably arise from the conversion of chloranil to trichlorhydroxyquinone (15). The writer finds that only reagent grade thymine and adenine gives these purple colours with the characteristic trichlorhydroxyquinone absorption at 520 nm. Pure thymine appears to have no reaction with chloranil in a spectroscopic grade dimethyl sulphoxide whereas pure adenine gives red coloured solutions as discussed below.

Machmer and Duchesne (16) and with Read (17) have shown that various purines and pyrimidines on mixing with chloranil in dimethyl sulphoxide give reddish coloured solutions. These colours take many hours to develop. The guanine chloranil solution took two weeks to reach the maximum colouration. This has been explained as due to the very slow formation of charge transfer complexes, the colouration being that of the charge transfer band. It is claimed that a correlation is found between the position of the band maximum and the electron donating ability of the purine or pyrimidine, as determined theoretically. Conversely the writer (18) has shown that whilst solutions do exhibit red colours on mixing in dimethyl sulphoxide, this is a secondary effect. Immediately on mixing, the solutions turn a golden yellow, displaying the spectrum of the chloranil anion, with absorption maxima at 315, 330, 427 and 455 nm. In time, a new peak begins to grow on the long wavelength side at about 520 nm. The resultant spectrum of the solutions is a composite of the two different spectra and the apparent peak heights and position depend on the time after mixing (see Fig. 4.6). It is believed that there is charge transfer immediately on mixing giving rise to the chloranil anion but that the later colouration is due to some chemical interaction involving the anion. This interpretation has been contested by Fulton and Lyons (19) who have produced a table of maxima of the new absorption peaks of a wide variety of purines and pyrimidines. They have shown that some compounds which according to Slifkin (18) were inactive with chloranil do interact on heating to give the

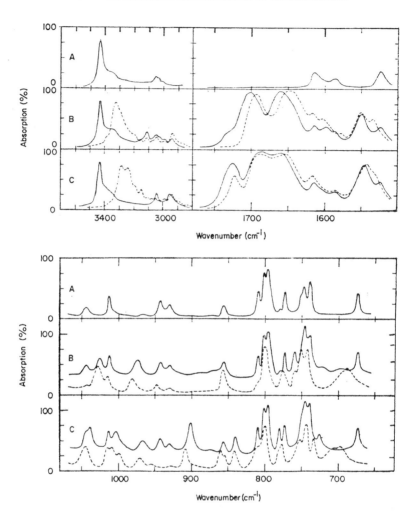

Fig. 4.5. Infra-red spectra as mulls in liquid paraffin or fully fluorinated oil in regions where differences are most marked between complexes and their components. (A) 3 : 4-5 : 6-Dibenzocarbazole ———. (B) 1 : 1 Mixture of 3 : 4-5 : 6-dibenzocarbazole and caffeine ———. Corresponding complex -----. (C) 1 : 1 Mixture of 3 : 4-5 : 6-dibenzo-carbazole and tetramethyluric acid ———. Corresponding complex -----. In B and C, from 1060 to 750 cm.$^{-1}$, the curve of the mixture is displaced upwards corresponding to 25% absorption. Reproduced with permission from Fig. 1, ref. 7.

red colour. This is attributed to an increase of the solubility of the purine or pyrimidine in the warmed solvent.

More recently, attempts have been made by the writer and Kushelevsky (to be published) to determine the thermodynamic parameters of these supposed complexes. Not only could no relationships be found between concentration and absorbance but no reversibility with temperature could be detected, a *sine qua non* for a weak complex. Attempts have also been made to obtain the infra-red spectrum of the red product. It proved impossible to isolate a solid from these solutions by evaporation under a low pressure, as one could only evaporate down to a thick black viscous fluid but no further. As chloranil itself forms charge transfer complexes with chloranil this is not so surprising (20). Additionally dimethyl sulphoxide has been shown to destroy chloranil on heating (21). It was found possible to isolate a red solid from solutions in 50% aqueous acetone similar in colour to those in dimethyl sulphoxide. The infra-red spectra of these solids showed no sign of chloranil and can only be interpreted as arising from the destruction of chloranil in solution.

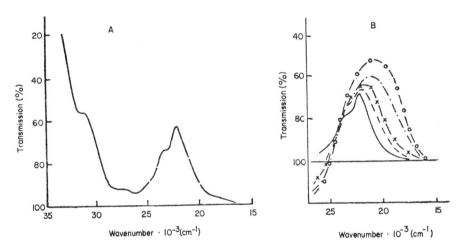

Fig. 4.6. Difference spectra. (A) Adenine and chloranil *vs.* chloranil in dimethyl sulphoxide. (B) Adenine and chloranil *vs.* chloranil in dimethyl sulphoxide immediately on mixing ———, after 15 min −−−−−, after 30 min — × — × —, after 45 min —·—·—, after 90 min — ○ — ○ —. Reproduced with permission from Fig. 1, ref. 18.

Certain purines and pyrimidines give green solutions and green solids whose infra-red spectra are simply the sum of the free components thus showing that not all purines or pyrimidines interact with chloranil and this is not just a question of solubility, as stated by Fulton and Lyons (19), as in the rotary evaporator very high concentrations indeed can be reached. One interesting result is with guanine. The infra-red spectrum of the solid complex shows the typical features of the quinhydrone type complex (22, 23). The carbonyl band of free chloranil at 1690 cm^{-1} is shifted in the complex to 1633 cm^{-1} and a new band appears at 1310 cm^{-1}. It would appear that charge transfer complexes are formed on initially mixing but that these rapidly go over to some red chemical compound. It will be recalled that the interaction between guanine and chloranil giving the red colour took two weeks to reach completion.

4.4. Acridine

Duchesne and Machmer (24) have shown that mixtures of guanidylic acid, thymidylic acid or DNA with acridines go yellow in dimethyl sulphoxide. No association constants have been derived nor has the reversibility of the changes been demonstrated. In view of the known reactivity of the solvent the designation of these colours as charge transfer bands is very speculative.

4.5. Azaguanine Self-complex

8-Azaguanine forms an unusual complex. MacIntyre (25) has shown that in the crystal molecules are associated in pairs at a separation of 3.25 Å, the typical separation in charge transfer complexes. Furthermore the two molecules forming a pair, are slightly displaced from exact super-position which would be necessary to permit the π-orbitals to overlap in order to form a charge transfer complex. No charge transfer band is observed but this would probably be in the far ultra-violet because of the low electron affinity of 8-azaguanine (49a).

4.6. Vitamin B_{12b}

The purines complex as donors with vitamin B_{12b}, in a similar manner to the amino acids (26). 8-Azaguanine form a remarkably strong complex having $\Delta H°$ of -27 kcal/mole. The behaviour of 8-azaguanine is not predicted by the molecular orbital calculations of the Pullmans which do not give either a low ionization potential or a high electron affinity (49a). However azaguanine does possess regions of high electron negativity due to the two adjacent nitrogens of the aza group and there may be some localized charge transfer interaction in complexing with the vitamin.

4.7. Haematoporphyrin

Purines form complexes with haematoporphyrin. Enthalpies of dissociation for the caffeine and 6-aminopurine complexes are -10.8 and -3.5 kcal/mole respectively (26, 27). The spectra are similar to those of amino acid complexes (see Section 3.5).

4.8. Iodine

Mixed solutions of many purines and pyrimidines with iodine in water display two effects (28). Initially on mixing, the spectrum of aqueous iodine which has absorption bands at 465 nm due to solvated iodine and at

353 and 287 nm due to I_3^-, changes with time, the I_2 band decreasing and the I_3^- band increasing with an isosbestic point. Similar spectral changes have been observed with other donors and interpreted as due to charge transfer (29). The secondary effect is the removal of the spectral bands associated with both the solvated iodine and the triiodide, and its replacement by a band in the far ultra-violet due to the formation of a purine or pyrimidine iodide. The schematic representation of this is the same as that shown for the interaction of the amino acids with iodine in Section 3.8.

4.9. Riboflavin

Several purines and pyrimidines, with the surprising exception of 8-aza-adenine, interact with the biological quinone riboflavin in water, as shown by the changes in absorption of the riboflavin. This is interpreted as due to the stabilizing of the semiquinone form of the flavin by charge transfer. This is discussed in some detail in Section 7.8.1.

The kinetics of the riboflavin tetramethyluric acid complex has been studied in some detail and it has been shown that a weak reversible complex is formed in water with an enthalpy of dissociation of -2.2 kcal/mole (30). See also Section 7.4.

4.10. Steroids

Apparent association constants have been evaluated for the interaction of purines and pyrimidines with steroids in dilute aqueous solution (31). These constants were obtained using a solvent partition method, which gives the ratio of bound steroid in the aqueous phase to the initial steroid concentration in cyclohexane. The interesting point is that these constants mirror the ionization potentials of the molecules, as derived from theoretical studies. A study of different steroids with one purine or pyrimidine shows that the formation constants also mirror the electron affinity of the steroids.† Deviations from the general relationships are explained in terms of steric hindrance. No correlation between the theoretical electron affinities of the purines or pyrimidines could be made for the interactions with a given steroid (see Table 4.6).

4.11. ESR and NMR Spectroscopy

In an attempt to observe charge transfer in various purine and pyrimidine complexes, electron spin resonance measurements have been made on a large number of the complexes already discussed (32). No signal corresponding to any major transfer of charge in complexes involving hydrocarbons or steroids was observed, contrary to the observations made on complexes between these molecules and iodine.

† Lata has pointed out in a private communication that as the steroids have different distribution coefficients they cannot be intercompared.

The nature of the interaction between the aromatic hydrocarbons benzene and toluene and the pyrimidines 1,3-dimethyluracil and 1,3-dimethylthymine has been studied by NMR. The equilibrium quotients of the four complexes have been determined from the signal shifts. These equilibrium quotients are the equilibrium or association constants of other workers. Rosenthal points out that the term equilibrium constant should be used for the product of the equilibrium quotient and the appropriate ratio of activity coefficients. It is concluded that the aromatic molecule associates with the pyrimidine ring in a vertical stacking arrangement, the intermolecular force being a charge transfer force with the pyrimidine as the electron acceptor (33) (see Table 4.5).

4.12. Tryptophan

Fluorescence studies on purine tryptophan complexes are discussed in Section 5.13.

4.13. DNA and RNA

In recent years, attention has been focussed on complexes involving DNA or RNA rather than their individual purine or pyrimidine bases.

4.13.1 AROMATIC HYDROCARBONS

Booth and Boyland (4) showed that DNA in the form of sodium deoxyribonucleate solubilized various polycyclic aromatic hydrocarbons in a similar manner to the bases, with the interesting difference that whereas the molar ratio of purine to hydrocarbon in the precipitated complex decreases with increasing concentration of purine, the molar ratio of DNA to the hydrocarbon increases with increasing concentration of hydrocarbon. The fluorescence maxima of the hydrocarbons are red-shifted and decreased in intensity as with the purines. The solubilizing effect of DNA on polycyclic aromatic hydrocarbons has also been reported by Liquori et al. (34) and T'so and Lu (35) who show that denaturation of the DNA, i.e. breaking down of the double standed form of helix into a single stranded molecule, increases the solubilizing effect.

Boyland and Green (36) have carried out similar observations on a wide range of hydrocarbons and obtained similar results. They interpret the interaction as being the intercalation of the hydrocarbon between the base pairs in the double helix of DNA.

This view is rejected by Giovanella, McKinney and Heidelberger (37) who have pointed out that there is not enough space between the base pairs for the hydrocarbons to intercalate. They believe that the solubilizing effect of DNA on these molecules is simply due to the formation of aqueous colloidal suspensions and have demonstrated this by producing very similar results using soap. However Boyland and Green together with Liu (38) have reaffirmed their view that there is a genuine solubilizing effect with DNA and have shown that the absorption spectra of the aqueous colloidal suspensions produced in soap solutions are slightly different from those produced in DNA solutions.

The effect of formaldehyde on the solubilization of the hydrocarbons by DNA would suggest that intercalation does occur because even in denatured DNA there is some strand recombination in solution. The apparent differences between the results of different workers is apparently due to the different ionic strength DNA solutions used (38).

Green and McCarter (39) have measured the polarized fluorescence of some polycyclic aromatic hydrocarbons in aqueous DNA solutions and examined the effect of flow orientation. Unlike the fluorescence of the hydrocarbons in DNA solution which is polarized, the fluorescence in caffeine solution is unpolarized. On flow streaming the DNA hydrocarbon solutions, there is observed a constant intensity of fluorescence which implies that the hydrocarbons are arrayed in a specific manner in the complex. These results are consistent with the intercalation theory though no light is shed as whether charge transfer forces are involved.

Ball, McCarter and Smith (40) have shown that the loss of absorbance of DNA solutions on heating follows the loss of absorbance of a DNA control solution. The viscosity of DNA and DNA : 3,4-benzopyrene solutions increase the range of temperature over which the loss of stability of the complexes occur. This is confirmation for the ideas of those authors that the polycyclic aromatic hydrocarbons form complexes with DNA rather than aqueous colloidal suspensions.

Solid gels of DNA under high vacuum show a DC conductivity which is greatly affected by the presence of aromatic hydrocarbons (41). By analogy with the work of Kearns, Tollin and Calvin (discussed in Section 6.2), in which the addition of electron acceptors to a porphyrin causes an increase in conductivity and is directly attributed to the presence of charge transfer states acting as a shallow trapping levels for the conduction electrons thus increasing the conductivity, it is suggested that the hydrocarbons behave as electron acceptors and DNA as an electron donor.

4.13.2. CATIONIC DYES

Electron paramagnetism has been demonstrated in complexes formed between DNA or RNA and cationic dyes when illuminated (42). It is assumed that electron transfer takes place from the dye to the nucleic acid. The spectra of the complexes in the form of thin films have been obtained and slight changes in the visible region as compared to uncomplexed dye noted. The intensity of the photoinduced paramagnetic signals are unaffected by the amount of naturation of the nucleic acid.

4.13.3. MUTAGENS

A considerable amount of study has been made on complexes of DNA with mutagenic agents such as acridine, proflavine and related compounds (43, 44).

The concensus of opinion seems to be that the mutagens are intercalated between the base pairs of the nucleic acid in the same way as the aromatic hydrocarbons. However although many of these carcinogens have been shown to be charge donors (45), and Duchesne *et al.* (17) have suggested that the interaction between the mutagens and purines is a charge transfer interaction, this is an isolated opinion. There appears to be sufficient difference between these mutagens and the hydrocarbons interaction with nucleic acids to ascribe the interaction to different mechanisms in each case. A modern review and study by Van Duuren, Goldschmidt and Seltzman (44) concludes that binding occurs to external sites rather than by intercalation and that the binding of the mutagens is essentially electrostatic.

4.14. Self-complexes of Nucleosides and Nucleotides

Basu and Greist (46) have claimed that some purine and pyrimidine nucleosides form self-charge transfer complexes as the spectra of very concentrated solutions exhibit new bands in the near ultra-violet and the blue, assigned as charge transfer bands. However similar bands have been observed by the writer for the same purine and pyrimidines bases which obey the Beer-Lambert law and show no temperature reversibility. It would appear that these bands are either very weak bands of the monomeric bases or are due to impurities.

On the other hand, a study of the luminescence of dinucleotides at 77°K, shows that red-shifting and broadening of the fluorescence spectra takes place as compared with the spectra of the mononucleotides (47). In aqueous frozen solution in the presence of salt, association takes place between the mononucleotides. This is manifested in the same changes of fluorescence as observed with the dinucleotides. The strength of the interaction goes in the order Py-Py < Pu-Py < Pu-Pu (where Py stands for pyrimidine and Pu for purine). This type of spectral change is similar to those occurring on exciplex and excimer formation which it has been suggested arise from charge transfer interaction (48). Certainly the order of interaction is consistent with the relative electron donor and acceptor properties of the purines and pyrimidines (49).

4.15 Kinetic Study

The effect of caffeine on the coupling reaction rate between β-naphthol and *p*-diazobenzenesulphonic acid has been examined by Overbeck *et al.* (50). The coupling rate was modified in the presence of caffeine and hence a reaction between the components separately with caffeine was looked for. From solubilizing studies it was found that a 1 : 1 complex was formed between β-naphthol and caffeine with an average association constant of 80.2 M^{-1} and a 1 : 1 complex with p-diazobenzenesulphonic acid with an association constant of 1.22 M^{-1}.

4.16. Mechanism of Purine and Pyrimidine Complexing

There is some evidence to suggest that the purines and pyrimidines act as electron donors in the presence of conventional charge acceptors. In the case of the complexes with hydrocarbons there are various difficulties. The majority of authors in the field ascribe the undoubted weak complexing to dispersion forces and dipolar interactions. A minority of workers have described these complexes as charge transfer complexes. There are similarities between these complexes and charge transfer complexes but this reflects the fact that both kinds of complexes are very weak as compared to chemical compounds. One argument put forward to show that they are not charge transfer complexes is that no charge transfer transitions are observed. This is not surprising for even if they were charge transfer complexes, due to the ionization potentials of both the hydrocarbons and the purines and pyrimidines (49, 51), being of the order of 8 to 9 eV and the electron affinities being of the order of 1 eV (49, 52) any charge transfer bands would appear in the ultra-violet region. As both the hydrocarbons and the purines or pyrimidines strongly absorb in this region, such bands would be difficult to observe.

The evidence presented by theoretical calculation is not conclusive as these are based on π-electron interaction. In Sections 7.4 and 7.8.5 it is suggested that 6-substituted amino purines behave as n-electron donors towards the flavins which is not considered in the theoretical calculations. The excited state spectra of these complexes are difficult to explain in terms of charge transfer. The infra-red spectra of the complexes show some similarities to those of hydrocarbon quinone complexes, notably the shift of the carbonyl band. So little is known about the infra-red spectroscopy of weak charge transfer complexes that little inference can be drawn from these results.

The crystal structure of the pyrene tetramethyluric acid complex is very similar to that of other charge transfer complexes, but the intermolecular distance of 3.4 Å is about 0.1 Å larger than that of typical charge transfer complexes (53).

Foster (54) in a discussion on the nature of these complexes states that whereas solubilization of a given hydrocarbon correlates with electron-donor ability of the purines, there is no correlation with the electron affinity of the hydrocarbon with a given purine and hence the question of charge transfer complexing is uncertain. Correlation is a property that can be evaluated statistically. The writer has evaluated the correlation coefficients for five

TABLE 4.3

Solubilization power of tetramethyluric acid and electron affinities of
aromatic hydrocarbons

Hydrocarbon	Solubilizing power[a]	Electron affinities†				
		b	c	d	e	f
1,2,3,4-dibenzopyrene	4.80		0.683			
anthanthracene	4.80				1.78	0.291
perylene	3.32	0.80	0.727	0.88	1.75	0.307
1,2-benzopyrene	3.25		0.41		1.22	0.497
3,4-benzopyrene	2.63		0.676		1.59	0.365
pyrene	2.05	0.68	0.417	0.57	1.23	0.445
1,2,5,6-dibenzanthracene	1.98	0.65	0.501	0.65	1.10	0.474
chrysene	1.88	0.04	0.313	0.47	0.98	0.520
1,2-benzanthracene	1.85	0.62		0.61	1.33	0.452
phenanthrene	1.76	− 0.20	0.014	0.17	0.69	0.605
anthracene	1.60	0.49	0.147	0.58	1.41	0.415

[a] From ref. 13, with permission.
[b] Hedges, R. M. and Matsen, F. A. (1958). *J. chem. Phys.* **28**, 950.
[c] Ehrenson, S. (1961). *J. phys. Chem.* **65**, 706, 712.
[d] Scott, D. R. and Becker, R. S. (1962). *J. phys. Chem.* **66**, 2713.
[e] From ref. 51a, with permission.
[f] Streitwieser, A. (1961). "Molecular Orbital Theory for Organic Chemists." Wiley, New York.
† Electron affinities in eV except f which is in molecular orbital coefficients.

TABLE 4.3a

Coefficients of correlation of linear plots of electron affinities of aromatic hydrocarbons *vs.* solubilization power of tetramethyluric acid

Confidence level	Coefficient of correlation	No of molecules	E_A values
85%	0.50	7	b
98%	0.70	9	c
96%	0.70	7	d
98%	0.68	10	e
98%	− 0.68	10	f

[a] Hedges, R. M. and Matsen, F. A. (1958). *J. chem. Phys.* **28**, 950.
[b] Ehrenson, S. (1961). *J. phys. Chem.* **66**, 706.
[c] Scott, D. R. and Becker, R. S. (1962). *J. phys. Chem.* **66**, 2713.
[d] Slifkin, M. A. (1963). *Nature*, **200**, 877.
[e] Streitwieser, A. (1961). "Molecular Orbital Theory for Organic Chemists." Wiley, New York.
Solubilizing powers from ref. 13.

different sets of electron affinities and found their significance using the "Student's" t distribution. The results are shown in Table 4.3. It is seen that one set is only significant at the 85% level, one is significant at the 96% level and three are significant at the 98% level. The coefficients of correlation of the latter four are about 0.7. This means that about half of the variation of solubilization can be attributed to a linear relationship between the two parameters with a high level of confidence.

In Table 4.4, it is shown that there is a high correlation $ca.$ -0.91, between the sum of the van der Waal's and dispersion forces and the solubilizing

TABLE 4.4

Correlation of solubilizing power of tetramethyluric acid with dispersion forces

Hydrocarbon	Interaction Energy[a]		Solubilizing Power[b]
	c	d	
1,2,3,4-dibenzopyrene	-11.33	-8.46	4.80
anthanthracene	-9.67	-6.20	4.80
perylene	-8.54	-5.76	3.32
1,2-benzopyrene	-9.74	-7.45	3.25
3,4-benzopyrene	-8.61	-6.40	2.63
pyrene	-7.73	-5.88	2.05
1,2,5,6-dibenzanthracene	-8.16	-6.23	1.98
chrysene	-7.54	-6.15	1.88
1,2-benzanthracene	-7.71	-5.75	1.85
phenanthrene	-6.76	-5.58	1.76
anthracene	-6.93	-5.22	1.60
coefficient of correlation	-0.91	-0.68	
confidence level	$\gg 99.5\%$	99.25%	

[a] Caillet, J. and Pullman, B. (1968). "Molecular Association in Biology," (Ed. B. Pullman). Academic Press, New York.
[b] From ref. 13, with permission.
[c] Total energy.
[d] Total energy less contribution from ionized forms.

TABLE 4.5

Equilibrium quotients of hydrocarbon pyrimidine complexes

Compounds	Equilibrium quotients
1,3-dimethyluracil, benzene	0·232
1,3-dimethyluracil, toluene	0·316
1,3-dimethylthymine, benzene	0·124
1,3-dimethylthymine, toluene	0·151

Reproduced with permission from Table II, ref. 33.

power of tetramethyluric acid, leaving about 17 % of the binding unexplained. The polarization forces can be expressed in the valence bond approach as the sum of different structures including those of a charge transfer nature (55).

TABLE 4.6

Interaction of steroids with pyrimidines $S_b/S_t \times 10^5 \pm$ s.d.

Steroid[a]	Pyrimidine[a]				
	Uracil	Thymine	Isocytosine	Cytosine	Alloxan
testosterone	3.7±0.7	2.0±0.4	3.4±0.4	1.8±0.1	1.7±0.4
epitestosterone	0.5±0.1	2.0±0.5	3.1±0.3	1.9±0.5	1.6±0.4
17α-methyltestosterone	2.3±0.7	1.5±0.6	3.4±0.7	0.5	0.8
testosterone acetate	(1.6±0.0)	(0.2)	(2.4±0.2)	(1.2±0.7)	(2.2±0.2)
testosterone propionate	(2.6±0.7)	(0.0)	(0.5)	(0.5)	(1.8±0.2)
progesterone	2.2±0.3	0.6	1.5±0.2	1.8±0.0	1.9±0.5
17α-hydroxyprogesterone	2.6±0.1	1.6±0.1	3.3±0.6	3.1±0.2	3.2±0.5
Δ^4-androsten-3,17-dione	5.0±0.5	1.4±0.0	0.7	2.0±0.2	1.6±0.4
Δ^4-androsten-3,11,17-trione	5.6±0.5	1.0±0.0	2.2±0.7	3.1±0.5	—
Δ^4-cholesten-3-one	0.5	0.5	0.7	0.2	0.4

[a] Steroid concentration: 2×10^{-5} to 6×10^{-5} M in stock cyclohexane solutions. Pyrimidine concentrations: 7.14×10^{-5} to 7.32×10^{-5} M; pH: 6·5–7.0.

Interaction of various steroids with some purine derivatives $S^b/S^t \times 10^4 \pm$ s.d.

Steroid[a]	Purine derivative[a]		
	Xanthine	Adenine	Adenosine
testosterone	3.2±0.1	3.0±0.3	2.5±0.5
epitestosterone	2.8±0.1	2.1±0.3	1.3±0.0
17α-methyltestosterone	0.6	1.9±1.1	1.0±0.1
progesterone	(3.4±0.1)	(2.7±0.8)	(3.8±0.0)
17α-hydroxyprogesterone	2.7±0.1	5.1±0.5	3.2±0.1
Δ^4-androsten-3,17-dione	2.8±0.1	2.4±0.3	5.8±0.1
Δ^4-androsten-3,11,17-trione	1.4±1.1	1.8±0.1	1.1±0.1

[a] Concentrations of interactants: steroids: 7.15×10^{-5} to 8.00×10^{-5} M; xanthine: 7.00×10^{-5}; adenine: 7.25×10^{-5}; adenosine: 8.00×10^{-5} M.
Reproduced with permission from Tables 1 and 2, ref. 31.

If we look only at those forces from which any dependence of ionized states of the molecules are removed we obtain a correlation coefficient of *ca.* -0.68, so that approximately half the variation can be ascribed to other than charge transfer forces and half to change transfer forces. Even if we take charge transfer forces as contributing to only 17 % of the binding of these complexes, these complexes would still be charge transfer complexes as formulated by Mulliken (Section 1.4). It is not a necessary condition of the Mulliken formulation that charge transfer forces be the major binding forces in charge transfer complexes.

4.17. Conclusion

The purines and pyrimidines form charge transfer complexes with organic acceptors although in some cases these may convert quite rapidly to chemical compounds. The weak complexes formed with aromatic hydrocarbons are stabilized in part by charge transfer forces.

REFERENCES

1. Brock, N., Druckrey, H. and Hamperl, H. (1938). *Arch. exp. Path. Pharm.* **189**, 709.
2. Weil-Malherbe, H. (1946). *Biochem. J.* **40**, 351.
3. De Santis, F., Giglio, E., Liquori, A. M. and Ripamonti, A. (1961). *Nature,* **191**, 900.
   Damiani, A., Giglio, E., Liquori, A. M., Puliti, R. and Ripamonti, A. (1966). *J. molec. Biol.* **20**, 211; (1967) **23**, 113.
4. Booth, J. and Boyland, E. (1953). *Biochim. biophys. Acta,* **12**, 75.
5. Booth, J., Boyland, E., Manson, D. and Wiltshire, G. H. (1951). *Rep. Brit. Emp. Cancer Campaign,* **29**, 27.
6. Weil-Malherbe, H. (1946). *Biochem. J.* **40**, 363.
7. Booth, J., Boyland, E. and Orr, S. F. D. (1954). *J. chem. Soc.* 598.
8. Boyland, E. and Green, B. (1962). *Brit. J. Cancer,* **16**, 347.
9. Van Duuren, B. L. (1964). *J. phys. Chem.* **68**, 2544.
10. Wentworth, W. E. and Becker, R. S. (1962). *J. Am. chem. Soc.* **84**, 4263.
11. Slifkin, M. A. and Walmsley, R. H. (1971). *Photochem. Photobiol.* **13**, 57.
12. Pullman, B. (1964). *Biopolymers,* **1**, 141.
   Caillet, J. and Pullman, B. (1968). "Molecular Associations in Biology" (Ed. B. Pullman). Academic Press, New York; and references therein.
13. Mold, J. D., Walker, T. B. and Veasey, L. G. (1963). *Anal. Chem.* **35**, 2071.
14. Beukers, R. and Szent-Györgyi, A. (1962). *Rec. Trav. Chim.* **81**, 541.
15. Slifkin, M. A., Sumner, R. A. and Heathcote, J. G. (1967). *Spectrochim. Acta,* **23A**, 1751
16. Machmer, P. and Duchesne, J. (1965). *Nature,* **206**, 618.
17. Duchesne, J., Machmer, P. and Read, M. (1965). *Cr. hebd. Séanc. Acad. Sci. Paris,* **260**, 2081.
18. Slifkin, M. A. (1965). *Biochim. biophys. Acta,* **103**, 365.
19. Fulton, A. and Lyons, L. E. (1968). *Austr. J. Chem.* **21**, 419.
20. Slifkin, M. A. (1969). *Spectrochim. Acta,* **25A**, 1037.
21. Slifkin, M. A. (1964). *Spectrochim. Acta,* **20**, 1543.
22. Slifkin, M. A. and Walmsley, R. H. (1970). *Spectrochim. Acta,* **26A**, 1237.
23. Slifkin, M. A. and Walmsley, R. H. (1969). *Experienta,* **25**, 930.
24. Duchesne, J. and Machmer, P. (1965). *C.r. hebd. Séanc. Acad. Sci. Paris,* **260**, 4279.

25. MacIntyre, W. M. (1965). *Science*, 147, 507.
26. Heathcote, J. G. and Slifkin, M. A. (1968). *Biochim. biophys. Acta*, 158, 167.
    Slifkin, M. A. and Heathcote, J. G. (1968). "Molecular Associations in Biology" (Ed. B. Pullman), p. 343. Academic Press, New York.
27. Heathcote, J. G., Hill, G. J., Rothwell, P. and Slifkin, M. A. (1968). *Biochim. biophys. Acta*, 153, 13.
28. Slifkin, M. A. (1965). *Biochim. biophys. Acta*, 103, 365.
29. Slifkin, M. A. (1965). *Spectrochim. Acta*, 21, 1391.
30. Slifkin, M. A. (1965). *Biochim. biophys. Acta*, 109, 617.
31. Molinari, G. and Lata, G. F. (1962). *Arch. Biochem. Biophys.* 96, 486.
32. Jones, J. B., Bersohn, M. and Neice, G. C. (1966). *Nature*, 211, 309.
33. Rosenthal, I. (1969). *Tetrahed. Lett.* 39, 3333.
34. Liquori, A. M., DeLerma, B., Ascoli, B., Botre, C. and Trasciatti, M. (1964). *J. molec. Biol.* 8, 20.
35. T'su, P. O. P. and Lu, P. (1964). *Proc. natn. Acad. Sci.* 51, 17.
36. Boyland, E. and Green, B. (1962). *Brit. J. Cancer*, 16, 507.
37. Giovanella, B. C., McKinney, L. E. and Heidelberger, C., (1964). *J. molec. Biol.* 8, 20.
38. Boyland, E. and Green, B. (1964). *Biochem. J.* 92, 4c.
    Boyland, E., Green, B. and Liu, S. L. (1964). *Biochim. biophys. Acta*, 87, 653.
39. Green, B. and McCarter, J. A. (1967). *J. molec. Biol.* 29, 447.
40. Ball, J. K., McCarter, J. A. and Smith, M. F. (1965). *Biochim. biophys. Acta*, 103, 275.
41. Snart, R. S. (1967). *Trans. Farad. Soc.* 63, 2384.
42. Balazs, E. A., Young, M. D. and Phillips, G. O. (1968). *Nature*, 219, 154.
43. Lerman, L. S. (1963). *Proc. natn Acad. Sci.* 49, 94.
44. Van Duuren, B. L., Goldschmidt, B. M. and Seltzman, H. H. (1969). *Annl. N.Y. Acad. Sci.* 153, 744 and references therein.
45. Briegleb, G. and Czekalla, J. (1959). *Z. Elektrochem.* 63, 6.
46. Basu, S. and Greist, J. H. (1963). *J. Chim. Phys.* 407.
47. Hélène, C. and Michelson, A. M. (1967). *Biochim. biophys. Acta*, 142, 12.
48. Slifkin, M. A. (1963). *Nature*, 200, 766.
    A comprehensive review of this topic is given by:
    Birks, J. B. (1970). "Photophysics." Wiley-Interscience, London.
49a. Pullman, B. and Pullman, A. (1963). "Quantum Biochemistry." Wiley-Interscience, New York.
49b. Berthod, H., Geissner-Prettre, C. and Pullman, A. (1966). *Theoret. Chim. Acta*, 5, 53.
50. Overbeek, J. Th. G., Wink, C. J. and Deenstra, H. (1955). *Rec. Trav. Chim.* 74, 69.
51a. Slifkin, M. A. (1963). *Nature*, 200, 877.
51b. Slifkin, M. A. and Allison, A. C. (1969). *Nature*, 215, 949.
52a. Lifschitz, C., Bergmann, E. D. and Pullman, B. (1967). *Tetrahed. Lett.* 46, 4583.
52b. Bergmann, E. D. (1968). "Molecular Associations in Biology" (Ed. B. Pullman). Academic Press, New York.
52c. Briegleb, G. (1964). *Angew. Chem.* 76, 326.
53. Wallwork, S. C. (1961). *J. chem. Soc.* 494.
54. Foster, R. (1969). "Organic Charge Transfer Complexes," p. 356. Academic Press, London.
55. Dewar, M. J. S. and Thompson, C. C. (1966). *Tetrahed. Supp.* 7, 97.

CHAPTER 5

Indoles and Imidazoles

5.1. Introduction

Indole itself whilst having little biological importance is the parent of such important compounds as tryptophan and serotonin (5-hydroxyl tryptamine).

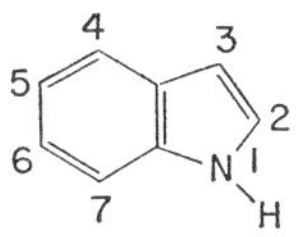

indole

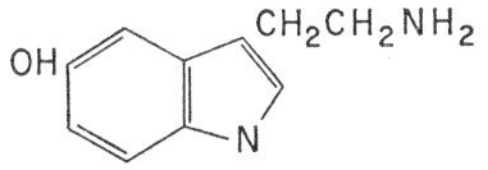

serotonin (5-hydroxyl-tryptamine)

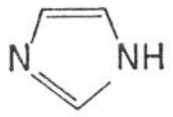

imidazole

Fig. 5.1. Structures of indole, serotonin and imidazole.

The first studies on the complexing ability of indole derivatives were those of tryptophan and serotonin with flavins (Section 7.2).

5.2. Chloranil

Indole gives coloured solutions in the presence of such well-known acceptors as trinitrobenzene, benzoquinone, chloranil and bromanil (1). A thorough study of the interaction of indole and methylindoles with chloranil in CCl_4 has been made by Foster and Hanson (2). They showed that indole forms a 1 : 1 charge transfer complex with chloranil with a charge transfer band whose maximum absorption appears at 496 nm. Beukers and Szent-Györgyi gives this maximum at 490 nm (3) whereas Nash and Allison puts it at 487 nm (4). The extinction coefficient is 1510 and the enthalpy of dissocia-

TABLE 5.1

Data for indole charge transfer complexes

Acceptor	K_c (M^{-1})	ΔH (kcal/mole)	ΔS (e.u.)	Temp (°C)	Solvent
tetracyanoethylene[a]	2.78	− 3.0	− 8.3	25	A
chloranil[b]	2.86	− 1.9	− 4.4	20	B
imidazolium perchlorate[c]	2.2[f]	− 1.8		22	C
imidazolium perchlorate[c]	1.6[g]	− 3.2		22	C
3-methyl imidazolium perchlorate[c]	1.8	− 3.2		22	C
histidinium perchlorate[c]	2.0[f]	− 1.8		22	C
histidinium perchlorate[c]	1.5[g]	− 3.2		22	C
imidazolium perchlorate[c]	2.0	− 3.0		22	D
histidinium perchlorate[c]	2.3	− 3.0		22	D
3-methyl imidazolium perchlorate[c]	2.4	− 3.1		22	D
3-methylhistidinium perchlorate[c]	2.2	− 3.1		22	D
3-methylhistidinium perchlorate[c]	1.4	− 3.4		22	C
NAD[d]	4.13			27	C
dinitrobenzene[e]	0.32			33.5	E
trinitrobenzene[e]	1.65			33.5	E

A = CH_2Cl_2, B = CCl_4, C = H_2O, D = 95% glycerol, E = $CHCl_3$.
[a] Ref. 2, [b] Ref. 5, [c] Ref. 21, [d] Ref. 26, [e] Ref. 31.
[f] Saturation Temperature Method.
[g] Distribution Method.

tion is 1.9 kcal/mole. Although a charge transfer complex is formed initially, a chemical reaction also occurs between the indole and chloranil. Foster and Hanson (2) further showed that charge transfer bands are obtained from mixed solutions of indoles with 2,3-dichloro 5,6-dicyano-p-benzoquinone, tettracyanoethylene, bromanil, 2,3 dicyano-p-benzoquinone, trichlorobenzo-quinone, 2,6-dichloro-p-benzoquinone, 2,5-dichloro-p-benzoquinone, chloro-p-benzoquinone, methyl-p-benzoquinone and sym-trinitrobenzene. In some

TABLE 5.2

Charge transfer complexes, energies of the highest occupied molecular orbitals and localized electronic indices of indoles

Derivative of indole	Maximum of charge-transfer band with chloranil $(cm^{-1} \times 10^{-3})$[a]	Energy of highest occupied molecular orbital (β units)		Reactivity index at C-3 of indole ring							
				Total charge in ground state		Frontier electron density[d]		Super-delocaliz-ability[d]		Approximate super-delocalizability[d]	
		Hyp.[b]	Ind.[c]	Hyp.	Ind.	Hyp.	Ind.	Hyp.	Ind.	Hyp.	Ind.
2,3-dimethyl	15.15	0.450	0.443	1.147	1.142	0.600	0.654	1.668	1.812	1.334	1.476
1,2-dimethyl	16.53	0.484	0.496	1.198	1.225	0.614	0.618	1.656	1.680	1.268	1.286
2,5-dimethyl	17.32	0.505	0.504	1.196	1.223	0.616	0.634	1.608	1.652	1.220	1.258
2-methyl	18.52	0.508	0.507	1.195	1.220	0.616	0.632	1.598	1.640	1.212	1.246
3-methyl	19.01	0.479	0.476	1.123	1.089	0.560	0.590	1.514	1.576	1.170	1.240
1-methyl	19.05	0.502	0.519	1.173	1.175	0.562	0.576	1.530	1.510	1.120	1.110
unsubstituted indole	19.92	0.534	0.534	1.170	1.170	0.570	0.570	1.466	1.466	1.068	1.068
coefficient of correlation[e]		0.81	0.78	~ 0	~ 0	-0.65	-0.86	-0.94	-0.95	-0.96	-0.92

The frontier electron is that occupying the highest filled orbital.

The superdelocalizability is defined as $2\sum_{j=1}^{m} c^2_{jr}/x_j$ where c_{jr} is the coefficient of the rth atomic orbital in the jth molecular orbital and x_j is the energy of the jth orbital in β units i.e. the energy $= \alpha \pm x_j\beta$. (See p. 6, Chapter 1.)

[a] Results supplied by Foster.
[b] The usual hyperconjugative-inductive model was used.
[c] The inductive model.
[d] For an electrophilic reaction.
[e] Calculated for linear relationship between charge transfer maximum and the different parameters.
Adapted with permission from Table 1 and Table 2, ref. 12.

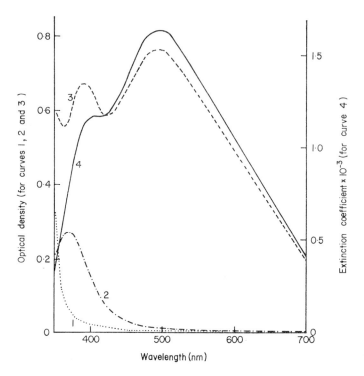

Fig. 5.2. Absorption spectra in carbon tetrachloride: (1) indole (0.233 M); (2) chloranil $(1.137 \times 10^{-3}$ M); (3) indole (0.233 M) + chloranil $(1.137 \times 10^{-3}$ M); (4) calculated absorption of the indole-chloranil complex alone. Reproduced with permission from Fig. 1, ref. 5.

unspecified complexes, double charge transfer bands have been observed. The appearance of two charge transfer bands is usually attributed to the donor molecule having two similar highest filled orbitals with slightly different ionization potentials. In the case of indoles this would not appear to be the explanation as the bands are only observed with some but not all the acceptors.

5.3. Iodine

Mixed solutions of indoles and iodine in dichloromethane have an absorption maximum at 367 nm but as the tri-iodide ion also absorbs in this region, no definite conclusion as to whether there is a charge transfer band have been made (2). In addition later irreversible changes also occur perhaps due to a chemical reaction between indole and iodine. Tryptophan has been shown to first complex with iodine and then to form an iodide (Section 3.8). In that case though tryptophan acts like the aliphatic amino acids as an n-donor.

5.4. Tetracyanoethylene

A 1 : 1 charge transfer complex is formed between indole and tetracyano-ethylene with a charge transfer maximum at 560 nm. The enthalpy of dissociation of the complex is 3.0 kcal/mole. There is however a later chemical reaction leading to the formation of 3-tricyanovinyl indole (5).

5.5. Organic Acceptors

Other electron acceptors which according to Szent-Györgyi *et al.* (6) form charge transfer complexes with indole are naphthoquinone, anthraquinone, maleic anhydride and trinitrobenzene.

The crystal structure of indoles and skatole complexes of 1,3,5-trinitrobenzene are very similar to those of other organic charge transfer complexes with an intermolecular spacing of 3.3 Å (7), i.e. characteristic of a charge transfer complex.

5.6. Oxygen

In the presence of oxygen at 110 atmospheres, solutions of indole in chloroform exhibit new absorption starting at 500 nm and increasing to shorter wavelength. This is interpreted as due to contact charge transfer between the indole as donor and the oxygen as acceptor. The ionization potential derived from this spectrum is 8.04 eV (8). Results for tryptophan are discussed in Section 3.2.

5.7. Pteridines

Fujimori (9) has investigated the interaction of tryptophan and serotonin with various pteridines. Solutions in phosphate buffer at pH 7 show slight colour due to absorption in the region of *ca.* 400 nm to 430 nm. Freezing such solutions gives an intensification of the colour, a good test of weak complexing. Equilibrium coefficients of the pteridine tryptophan and pteridine serotonin complexes have been evaluated spectrophotometrically. A tryptophan-containing protein bovine serum albumin complexes in a similar manner (see Table 5.3).

5.8. Acridine

Several authors have examined the interaction of the heterocyclic molecule acridine with indoles. Fujimori (9) has derived spectrophotometrically equilibrium constants for acridine tryptophan and acridine serotonin complexes. These are 45 M^{-1} and 60 M^{-1} respectively. Szent-Györgyi and McLaughlin (10) have shown that the colours produced when a wide variety of compounds mainly carcinogens are evaporated together with acridine on filter paper are the same. This, they suggest is due to the spectrum of acridine

TABLE 5.3

Charge transfer complexes of pteridines and tryptophan

	K_c	Tryptophan $\varepsilon_c - \varepsilon_p$	λ_{max}	K_c	Serotonin $\varepsilon_c - \varepsilon_p$	λ_{max}
amethopterin	45	1.6×10^2 (420) nm		60	1.6×10^2 (420) nm	
aminopterin	25	5.2×10^2 (430) nm		45	5.2×10^2 (430) nm	
N^{10}-methyl folic acid	22	3.6×10^2 (420) nm		34	3.6×10^2 (420) nm	
folic acid	13	1.1×10^3 (400) nm		30	1.1×10^3 (400) nm	
N^{10}-formyl folic acid	15	7.1×10^2 (400) nm		23	7.1×10^2 (400) nm	
9-methyl folic acid	9	8.5×10^2 (400) nm		20	8.5×10^2 (400) nm	
2,4-diamino-6,7-dimethyl-pteridine	7	-1.2×10^3 (370) nm		18	-1.2×10^3 (370) nm	
2-amino-4-hydroxy-pteridine-6-carboxylic acid	5	2.2×10^3 (410) nm		15	2.2×10^3 (410) nm	
xanthopterin	2	5.0×10^3 (410) nm		15	-7.7×10^3 (410) nm	
acridine	23	6.7×10^2 (390) nm		100	-1.1×10^3 (390) nm	

ε_c = extinction coefficient of complex.
ε_p = extinction coefficient of pteridine.
K_c in M^{-1}.
Reproduced with permission from Table 1, ref. 9.

being perturbed rather than to the appearance of charge transfer bands, as implied by Fujimori (8). Allison and Nash interpret the new bands seen in acridine complexes as arising from an n-π interaction (4).

5.9. Charge Complimentarity

Szent-Györgyi and Isenberg (11) have isolated a black solid from indole iodine mixtures which gives a large ESR signal and attributed this to the formation of a charge transfer complex.† Similar results were obtained with indole acetic acid and serotonin. It is suggested that the apparent strong charge donor ability of the indoles arises from the presence of strong negatively charged carbon atoms and that the charge transfer is localized at these atoms rather than arising from the donation of a π-electron of the conjugated electron system in the indole ring. The evidence for this is that carbazole which contains two π-electron conjugated systems but no negatively charged carbon atoms does not interact with iodine. Carbazole does form π-π complexes with chloranil in chloroform (unpublished work by the writer). Support for this view comes from the theoretical studies of Green and Malrieu (12). They have calculated the superdelocalizability at C_3 in the indole nucleus and found that this correlates better with the maximum position of the charge transfer band than it does with the energy of the highest

† A discussion of ESR signals in charge transfer complexes is given in Section 2.1.8.1.

occupied molecular orbital of the indole, a measure of the ionization potential (Table 5.2). Similar ideas have been put forward to explain the strong interaction of flavins with indoles (Chapter 6) and of trinitrobenzene (Section 5.18).

5.10. ESR Measurements of Iodine and Chloranil Complexes

Smaller and his colleagues (13) like Szent-György and Isenberg (11) have shown that indole and iodine produce a black complex which in the solid

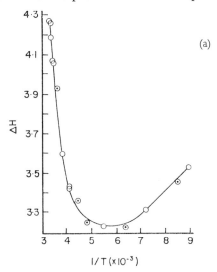

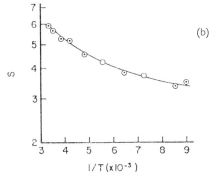

Fig. 5.3. Temperature dependence of conductivity and signal strength of indole-iodine complex.

(a) Line-width in oersteds vs. inverse temperature in inverse degrees K. The line-width is taken as the separation of the points of maximum slope of the absorption.

(b) Number of spins in 10^{19} spins/gm. vs. inverse temperature.

Reproduced with permission from Figs 1 and 2, ref. 13.

state give strong ESR signals. This complex is also a semi-conductor with a high conductivity and an activation energy of 0.11 eV. There is no correlation between the temperature dependence of the conductivity and the strength of the ESR signal which means that the charge carriers giving rise to the conductivity are not the same entities as give rise to the ESR signal. Consequently the ESR signal is not explained. In general though a marked increase in conductivity on complexing is a usual accompaniment to charge transfer (see Section 2.4).

Lu Valle and his coworkers have looked at the ESR signals of indole chloranil complexes (14). Two signals are observed, a broad peak similar to

TABLE 5.4

Relative ESR signal strength of some complexes

Acceptor	[2-(1-ethyl-3,3-dimethyl-indolylidene)]-[5-(1,3-diethyl-2-thiobarbituryl-idene)]-dimethine merocyanine	Electron Affinity (eV)[a]		
maleic anhydride	330	0.11	1.5	2.0
chloranil	100	1.37		
iodine	50			1.7
tetrachlorophthalic anhydride	50	0.56	1.8	
trinitroresorcinol	35			
dibromomaleic acid	20			
bromanil	12	1.37		
diaminoanthraquinone	10			
acenaphthequinone	11			
tetrabromophthalic anhydride	5			

[a] Electron affinities in left-hand column from Briegleb, G. (1964). *Ang. Chem.* **76**, 326. Other values from Batley, M. and Lyons, L. E. (1962), *Nature*, **196**, 573; and for iodine from Person, W. B. (1963), *J. chem. Phys.* **38**, 109.
Adapted with permission from Table 3, ref. 14.

those seen by the earlier workers, thought to arise from an oxidation-reduction i.e. complete charge transfer process, and in addition a sharp peak invariant with the different donors used provisionally called the donor acceptor peak, Fig. 2.3.

5.11. *N*-Methylphenazonium Sulphate

Adding indole or imidazole to *N*-methylphenazonium sulphate (PMS) in dimethyl sulphoxide produces a red colour. This coloured solution gives a very strong ESR signal (15), which is interpreted as due to the formation of the PMS free radical by charge donation from the indole or imidazole. The great similarity between the interaction of the two molecules with PMS not withstanding the indole having twice as many π-electrons as imidazole is interpreted to mean that the donated electron is a lone-pair electron from a nitrogen atom in the indole or imidazole ring.

5.12. Thiamine

A series of studies on the complexes formed between thiamine and indole derivatives have been carried out by Sable, Bigelow and others (16, 17, 18). It has been established that 1 : 1 complexes are formed between thiamine or thyamine pyrophosphate and tryptophan and other indoles using NMR,

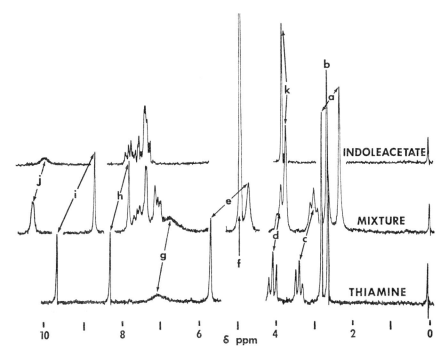

Fig. 5.4. NMR spectra of indoleacetate, thiamine and their mixture. Reproduced with permission from Fig. 3, ref. 17.

proton magnetic resonance and ultra-violet spectroscopy. Little interaction has been detected between thiamine and other aromatic amino acids. Association constants of these complexes are similar in magnitude to those obtained for typical charge transfer complexes in solution, and it is therefore believed that these are charge transfer complexes. Although no charge transfer bands

TABLE 5.5

Data of indole thiamine complexes

Effect of different salts on thiamine indoleacetate complex		
	Salt	K_c
		M^{-1}
A. [Thiamine] varied (0.1–0.5 M)	LiCl	2.9 ± 0.1
	NaCl	3.6 ± 0.2
	KCl	3.1 ± 0.1
	NH_4Cl	3.3 ± 0.1
	$(C_2H_5)_4NCl$	1.2 ± 0.1
B. [Indoleacetate] varied (0.1–0.5 M)	LiCl	1.1 ± 0.1
	NaCl	1.7 ± 0.1
	Sodium formate	1.7 ± 0.1
	KCl	1.4 ± 0.1
	KNO_3	1.5 ± 0.1
	$(C_2H_5)_4NCl$	1.4 ± 0.1

Apparent association constants of indolyl-thiamine complexes		
Acceptor	Donor	K_c
		M^{-1}
Thiamine (Va) (0.1–0.5 M)	5-hydroxytroptophan	6.3 ± 0.3
	indoleacetate	3.6 ± 0.2
	tryptophan	4.0 ± 0.1
	tryptamine	3.8 ± 0.3
	tryptophol	4.4 ± 0.3
Thiamine-PP (Vb) (0.081–0.296 M)	5-hydroxytryptophan	4.3 ± 0.2
	tryptophan	2.7 ± 0.1
	indoleacetate	2.9 ± 0.2
	tryptophol	3.5 ± 0.3
	tryptamine	3.4 ± 0.2

Thermodynamic constants for indoleacetate-thiamine complex				
Ionic strength	$K_c^{25°}$	$\Delta G^{25°}$	ΔH	$\Delta S^{25°}$
M	M^{-1}	kcal/mol	kcal/mol	cal/mol deg
1.0	3.6 ± 0.2	-0.76 ± 0.03	-2.5 ± 0.4	-5.9 ± 1.4

Reproduced with permission from Tables 3, 4 and 5, ref. 18.

as such have been detected, spectral changes on complexing are similar to those exhibited by other systems believed to be charge transfer complexes. The paramagnetic resonance signals show shifts which are taken to be indicative of π-π interaction. It was also noticed that the association constants of those complexes containing electron donating substituents were greater than those which did not, again indicative of charge transfer interaction. Association constants decrease with increasing ionic strength which would suggest that there is little ionic character in the ground state,

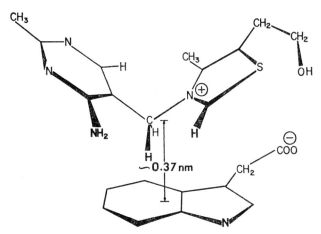

Fig. 5.5. Model of indoleacetate thiamine complex. Reproduced with permission from ref. 17.

although the authors in fact interpret this in just the opposite way. The structure of the complexes have been deduced primarily from the microwave data the structure of thiamine indoleacetate is shown in Fig. 5.5.

5.13. Nucleic Acids

An interesting interaction between tryptophan and various nucleic acids has been discovered by Montenay-Garestier and Hélène (19). Aqueous frozen solutions of the nucleic acids mixed with tryptophan have fluorescence spectra that lie well to the longwave length side of the fluorescence of the individual components. This long wavelength shift reaches a maximum in equimolar solutions, indicating the probable existence of a 1 : 1 complex. The spectrum obtained by reflection from the frozen solutions also lie to longer wavelength than those of the individual components. Concentration dependence absorption studies again indicate a 1 : 1 stoichiometry. It is concluded that charge transfer complexes are formed with tryptophan as the electron donor. Theoretical studies support the conclusion that the purines and pyrimidines may act as charge acceptors with the pyrimidines being better acceptors, which fits the observation that the transition energies of the pyrimidine complexes are lower than for the purine complexes. No interaction could be detected for the weaker donors, tyrosin and phenylalanine.

5.14. Purines

The writer (20) has examined the interaction of tryptophan with adenine, a nucleic acid base, in ice and found similar spectra to those reported for adenosine (19). The excitation spectrum of this adenine tryptophan complex is

TABLE 5.6

Data of the emission of adenine, tryptophan and their mixture

Compound	Excitation wavelength (nm)	Emission wavelength (nm)	
tryptophan	313	450	in ice at 77°K
adenine	317	341, 459	in ice at 77°K
mixture	303, 354	395	in KBr at 293°K
mixture	362	400, 445	in ice at 77°K

different to the excitation spectra of either adenine or tryptophan. The complex has also been studied at room temperature by evaporating down adenine tryptophan mixtures and dispersing the resulting yellow solid in KBr discs. The excitation and fluorescence spectra are similar to those of frozen aqueous solution. The infra-red spectrum of the complex in KBr has been measured and is consistent with the formation of a charge transfer complex, the tryptophan acting as donor. The primary features of the infra-red spectrum are slight shifts of the bands associated with the indole ring. This is consistent with the tryptophan being a π-electron donor but not an n-electron donor which would require that the tryptophan be in its $-NH_2$ form (see Section 3.3). Some changes have also been observed in the infra-red spectrum of an adenine phenylalanine mixture evaporated down and dispersed in KBr. These changes consist in the loss of several bands thought to arise from the $C=N$ and $C=C$ stretching modes of adenine at 1253 and 1602 cm^{-1} respectively. The complex possesses two new peaks at 792 and 1262 cm$^-$. Although Montenay-Garestier and Hélène (19) did not observe any fluorescence from phenylalanine adenosine mixtures, the possibility of such a complex existing is real.

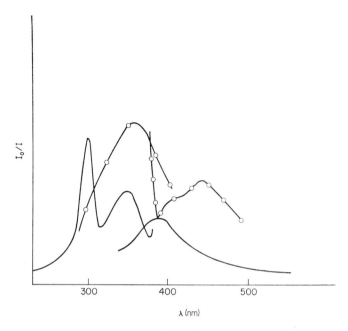

Fig. 5.6. Fluorescence and emission spectra of adenine tryptophan mixtures in KBr at room temperature ———; ice at 77°K — ○ — ○ —.

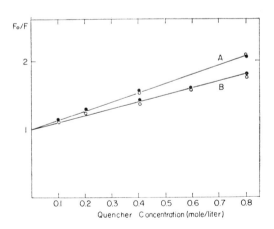

Fig. 5.7. Effect of histamine (curve A) and of imidazole (curve B) on the fluorescence intensity of lysozyme at 340 mμ. F_0, Fluorescence of lysozyme in the absence of quencher; F, fluorescence of lysozyme in presence of quencher. Lysozyme concentration, 80 μg/ml; acetate buffer of pH 5; temperature, 22°; ●, in water; ○, in glycerol-water (1 : 1, v/v). Excitation wavelength 290 nm. Reproduced with permission from Fig. 5, ref. 21.

5.15. Lysozyme

The interaction of several imidazole and indole derivatives with the protein lysozyme has been studied by Shinitzky et al. (21). The inhibition of lysozyme by the various molecules is first order. From temperature dependence studies of the equilibrium constants, binding energies have been obtained all of which are about $3 \cdot 0$ kcal M^{-1}. These inhibitors also quenches the fluorescence of the protein which contains many fluorescent tryptophyl residues. As it is the tryptophyl fluorescence which is quenched, this suggests that complexing takes place between the indoles or imidazoles and these residues in the protein.

5.16. Imidazole and Imidazolium

To test the hypothesis in Section 5.15, the authors (21) have examined the indole imidazolium interaction. Fluorescence quenching of the indole by the imidazolium was studied. The increased solubility of indole in water in the presence of imidazolium salts as a function of temperature was studied, it being assumed that any increase in solubility arose from the formation of water soluble indole-imidazolium charge transfer complexes. All the values of enthalpies of dissociation obtained from both techniques were in the range 1.4 to 3.4 kcal/mole. However the fluorescence quenching and solubility measurements using a phase distribution method both yielded the same value of 3.2 kcal/mole. No charge transfer bands were observed in absorption, but it was seen that the difference spectrum between the imidazolium and indole in separate cuvettes as compared to the mixture showed a loss of absorbance at 255 nm and a gain at 290 nm, both in the region of strong indole absorbance.

Solutions of imidazolium and histidinium iodide do not obey the Beer-Lambert law. In polar solvents an isosbestic point occurs at 248 nm. This is interpreted as arising from the formation of charge transfer ion pairs with the iodide ion as the donor, similar to the charge transfer ion pairs formed by the

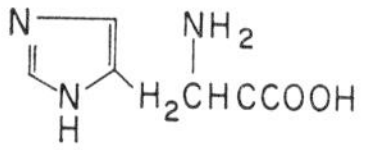

Fig. 5.8. Histidine.

pyridinium salts (Section 8.2). These experiments lead to the conclusion that the tryptophyl residues on the protein lysozyme form charge transfer complexes with indoles and imidazole, the residues being the charge donors.

The charge transfer complexes formed between indole and imidazole derivatives have been characterized (22) by an indirect method described by Smith (23). The addition of a charge donor to a solution already containing a charge transfer complex, causes a redistribution of the acceptor between the two donors. The decrease of the pre-existing complex is due to the formation of a new complex with the added donor. The kinetics of formation of the new complex can be obtained by following the kinetics of dissociation of the first complex. Association constants found by this method are somewhat lower than those found by the other methods (21).

Studies of intra-molecular indole-imidazolium complexes have been undertaken using the fluorimetric titration technique in aqueous solution (24). Fluorimetric titration consists of plotting the absolute quantum yield of fluorescence against pH and hence finding the pK value, which is that pH at which the fluorescence yield is halfway between its minimum and maximum values. Both natural peptides and synthetic peptides containing histidyl and tryptophy residues were studied. The fluorimetric titration curves of the model peptides coincided with the potentiometric titration curves of the imidazole side chain of the histidyl residue. Similar fluorimetric titration curves observed for natural peptides can be attributed to intramolecular complexing between histidyl and tryptophyl residues. Optical rotatory dispersion measurements give no information about the changes in protein conformation during changes in pH. The model compounds do however exhibit ORD changes in the far ultra-violet. A hyperchromic effect in the region of 240 to 250 nm is thought to be due to the characteristic charge transfer band of the indole imidazole complex.

Imidazole imidazolium mixtures obey the Beer-Lambert law (25). However an intramolecular charge transfer interaction has been suggested as occurring between imidazole and imidazolium both in poly N-vinyl imidazole and the dipeptide histidyl-histidine. The spectrophotometric titration curve has a minimum at pH 5, indicative perhaps of the intramolecular charge transfer. The lack of a charge transfer band in absorption is explained away by the weak electron donor ability of histidine which would put the band below 200 nm, i.e. in the quartz ultra-violet region.

5.17. NAD and Coenzyme Models

There have been many studies on complexes of NAD and other coenzymes. These are in the main discussed in Section 8.4. Interaction involving indoles and imidazoles are discussed here.

Alvisatos et al. (26) have looked at the nonenzymatic interaction of various indoles with NAD. The addition of indole to NAD solutions causes the appearance of a bright colour due to a marked increase of absorption in the 400 to 500 nm region. This colour shows a pH independence increasing markedly at high pH. Association constants and extinction coefficients have been evaluated at pH 2 using the concentration dependence of this new band.

TABLE 5.7

Data of indole NAD complexes

Compound	$K_c\,(\mathrm{M}^{-1})$	$\varepsilon_{380}\,(\mathrm{cm}^2/\mathrm{mole})$	pH range
serotonin[a]	11.10	367	2.0–2.4
tryptamine	14.55	156	1.9–2.2
tryptophan	13.40	151	1.9–2.5
indole	4.13	430	1.9–2.4

[a] Utilized as the creatinine sulfate complex.
Measured at 27°.
Reproduced with permission from Table 1, ref. 25.

The order of these constants is tryptamine > tryptophan > serotonin > indole, which is a different order to the strength of interaction of the indoles with FMN (Section 7.2). Freezing the indole NAD complex gives an increase in depth of colour. The spectral data indicates clearly a 1 : 1 stoichiometry, at least in solution. By analogy with the interactions with other molecules already discussed in this chapter it is presumed that these complexes are charge transfer complexes. It is recognized that other factors must be contributing to the complex stability as the order of complexing is not as predicted on the grounds of charge transfer alone. The site of the electron acceptance of the coenzyme is thought to be primarily on the nicotinamide moiety of NAD.

Cilento and coworkers (27, 28) have also examined the interaction of some indoles with NAD and confirm the spectral changes reported by Alivisatos et al. (26). Measurements have been made of the association constants of some indoles with 1-benzyl-3-carboxymide pyridinium chloride used as a model for oxidized pyridine coenzymes. The association constants are in a different order from those just quoted previously. It is not thought that the interactions are any different but it illustrates that many factors are involved in complexing even when similar molecules are being compared. An interesting suggestion made by Cilento et al. is that complex formation with tryptophan could be used for detecting exposed tryptophyl residues in proteins, cf. Swinehart and Hess's method for the detection of tryptophyl residues by complexing with RFN (Section 7.2).

A series of model compounds in which indoles have been directly incorporated into pyridinium chlorides have been synthesized and studied by Shifrin (29). The spectrum of indolylethylnicotinamide in methanol is different from the spectrum of a 1 : 1 mixture of tryptamine hydrochloride and nicotinamide methochloride. There is an increase in absorption in the region of 320 nm in the model compound thought to be a charge transfer

TABLE 5.8

Data for complexes of 1-benzyl-3-carboxamide pyridinium chloride
with indole and derivatives[a]

	Conditions	Association constant (M^{-1})
indole	water	2.21
	water	2.47
L-tryptophan	water	2.52
	water	2.20
yohimbine	pH 4.0[b]	1.42
	pH 2.2[b]	1.41
acetyltryptophan	pH 6.5	5.00
	pH 6.5	4.73
glycyl-L-tryptophan	pH 6.6	2.98
	pH 6.1	2.92
indole-3-acetic acid	pH 6.7	4.09
	pH 6.3	4.41
serotonin	pH 6.4	1.81
	pH 6.6	2.07

[a] Room temperature ($25° \pm 2°$).
[b] Aqueous HCl. All the other pH values refer to 1.7×10^{-2} M phosphate.
Reproduced with permission from Table 1, ref. 28.

band. If this model compound is reduced to indolylethyldihydronicotinamide
then the absorption spectrum of this molecule is identical to the spectrum of
a mixture of tryptamine hydrochloride and nicotinamide ethochloride.
Fluorescence studies of these compounds produced some interesting results.

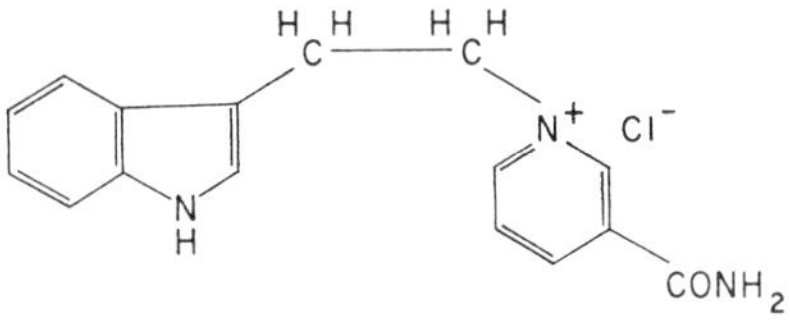

Fig. 5.9. N-(β-indolylethyl)-3-carboxamide pyridinium chloride.

When indolylethyldihydronicotinamide is excited in the region of the indole
absorption, the fluorescence spectrum of the molecule is that of indolylethyldihydronicotinamide, showing that the excitation energy is very efficiently
transferred from the indole moiety. Indolylethylnicotinamide on the other
hand shows no emission from the indole moiety nor is there any energy

TABLE 5.9

Association constants, K, of the tryptophan residues of lysozyme with various imidazole derivatives

Quencher	K litres/mole
imidazole	1.0 ± 0.1
L-histidine	0.9 ± 0.1
L-histidine methyl ester	1.3 ± 0.1
histamine	1.3 ± 0.1

Reproduced with permission from Table 3, ref. 21.

transfer to the nicotinamide. The new absorption in the model compound is attributed to intramolecular charge transfer by reason of the indoles being known charge donors and the pyridinium ring having good acceptor properties.

Other model compounds studied by Shifrin (30) are N-β-4'-imidazolylethyl)-3-carbamoylpyridinium chloride (ImCP) and N-(β-indolylethyl)-3-carbamoylpyridinium chloride (ICP). The absorption spectrum of ICP is very similar to that of indolylethylnicotinamide. The absorption spectrum of ImCP has a long wavelength band not exhibited by imidazole or the pyridinium ion which is assigned to an n-π charge transfer interaction between the imidazole as donor and the pyridinium as acceptor. Protonation of the imidazole, which blocks the lone-pair, i.e. n-electrons removes this new band. The dihydronicotinamide produced by dithionate reduction of ImCP shows no evidence of a charge transfer interaction neither in absorption nor in emission. Reduction causes removal of the positive charge of the pyridinium moiety thus removing the charge accepting ability.

5.18. NMR Studies

Complexes of indole and derivatives with the organic acceptors trinitrobenzene and dinitrobenzene have been studied by NMR (31). The strength of interaction, Table 5.10, correlates to some extent with the electron donating ability of the indole. Methylation, especially at the 3-position increases the association constant. However large substituents decrease the association constant suggesting that steric hindrance occurs. A detailed study of the chemical shifts on complexing shows that the hydrogens at the 2 and 3 positions of the indole are the most sensitive to complex formation, suggesting that complexing primarily arises from negative charge at the C_3 position, cf. Section 5.9.

TABLE 5.10

Association constants (K_c) in kg of solution per mole, maximum observed chemical shift (Δ_{max}), and chemical shift in the pure complex (Δ_0) relative to the shift of protons in unassociated acceptor molecules for 1,4-dinitrobenzene and 1,3,5-trinitrobenzene complexes with various donors at 33.5° in chloroform

Donor	1,3,5-trinitrobenzene			1,4-dinitrobenzene		
	$K_c{}^a$ (kg/mole)	Δ_0 (c/sec)	Δ_{max} (c/sec)	$K_c{}^a$ (kg/mole)	Δ_0 (c/sec)	Δ_{max} (c/sec)
3-methylindole	2.4_0	$97._7$	$70._5$	0.47	$115._7$	$44._0$
2-methylindole	2.1_6	$98._6$	$75._4$	0.41	$122._6$	$43._2$
7-methylindole	1.9_7	$99._4$	$73._2$	0.31	$152._6$	$46._6$
3-ethylindole	1.7_8	$102._0$	$72._7$	0.32	$148._3$	$40._3$
indole	1.6_5	$99._4$	$72._0$	0.32	$135._4$	$37._8$
3-dimethylaminomethylindole	1.3_4	$84._0$	$37._0$	0.26	$139._9$	$18._3$
3-diethylaminomethylindole	b			0.24	$148._3$	$22._6$
benzothiophen	0.8_9	$98._6$	$53._1$	0.24	$140._6$	$33._7$
thiophen	0.1_5	$127._7$	$19._2$	c		
pyrrole	0.2_5	$69._4$	$27._1$	c		

^{a} Estimated error $\pm 5\%$.
^{b} Could not be measured because of chemical reaction.
^{c} Too weak to measure.
Reproduced with permission from Table 1, ref. 31.

TABLE 5.11

Association constant (K_c) for the complex of indole with 1,3,5-trinitrobenzene in 1,2-dichloroethane at 33.5°

Proton measured	K_c (kg/mole)	Δ_0 (c/sec)	Δ_{max} (c/sec)
H_3 in indole	0.9_3	$43._0$	$24._0$
H_{arom} in indole^a	1.0_0	$18._6$	$10._1$
H in 1,3,5-trinitrobenzene	0.8_5	$96._6$	$57._3$

^{a} Measured for band of maximum absorption.
Reproduced with permission from Table 3, ref. 31.

5.19. Flavins

The interactions of indoles with flavins is discussed in Section 7.2 and with methylviologen in Section 9.8.

5.20. Serotonin Complexes

Solutions of serotonin and chloranil exhibit a new broad band in the region of 520 nm (32). The association constant derived from the Benesi-Hildebrand equation is 172×10^{-6} M^{-1}.

Differential spectrophotometry has revealed the existence of a complex between serotonin and diazapam (7-chloro-1,3-dihydro-1-methyl-5-phenyl-2H-1,4-benzdiazepin-2-one) with an association constant of 12×10^3 M^{-1}(33).

5.21. Conclusions

From all the evidence it is well established that the indoles form charge transfer complexes as donors. Imidazoles act as donors in the neutral normal form but can act as acceptors when protonated. Various modes of donation have been suggested for indole, π-donation, n-donation and charge complimentarity or localized donation form the C_3 position. Probably all these do occur under the appropriate conditions and the different explanations correct in their context. Tryptophan which contains both the indole ring and an amino group can in the right conditions act also as an n-electron donor from the nitrogen of the amino group as it does when complexing with chloranil (Section 3.2).

REFERENCES

1. Fujimori, E., quoted in ref. 10.
2. Foster, R. and Hanson, P. (1965). *Tetrahed.* **21**, 255.
3. Beukers, R. and Szent-Györgyi, A. (1961). *Rec. Trav. Chim.* **81**, 255.
4. Allison, A. C. and Nash, T. (1963). *Nature*, **197**, 758.
5. Foster, R. and Hanson, P. (1964). *Trans. Farad. Soc.* **60**, 2189.
6. Szent-Györgyi, A., Isenberg, I. and McLaughlin, J. (1961). *Proc. natn. Acad. Sci.* **47**, 1089.
7. Hanson, A. W. (1964). *Acta Crystallog.* **17**, 559.
8. Slifkin, M. A. and Allison, A. C. (1967). *Nature*, **215**, 949.
9. Fujimori, E. (1959). *Proc. natn. Acad. Sci.* **45**, 133.
10. Szent-Györgyi, A. and McLaughlin, J. (1961). *Proc. natn. Acad. Sci.* **47**, 1397.
11. Szent-Györgyi, A. and Isenberg, I. (1960). *Proc. natn. Acad. Sci.* **46**, 1334.
12. Green, J. P. and Malrieu, J. P. (1965). *Proc. natn. Acad. Sci.* **54**, 659.
13. Smaller, B., Isenberg, I and Baird, S. L. (1961). *Nature*, **191**, 168.
14. LuValle, J., Leiffer, A., Koral, M. and Collins, M. (1963). *J. phys. Chem.* **67**, 2636.
15. Kimura, J. E. and Szent-Györgyi, A. (1969). *Proc. natn. Acad. Sci.* **62**, 286.
16. Sable, H. Z. and Biaglow, J. E. (1965). *Proc. natn. Acad. Sci.* **54**, 808.

17. Biaglow, J. E., Mieyal, J. J., Suchy, J. and Sable, H. Z. (1968). *J. biol. Chem.* **244,** 4052.
18. Mieyal, J. J., Suchy, J., Biaglow, J. E. and Sable, H. Z. (1968). *J. biol. Chem.* **244,** 4063.
19. Montenay-Garestier, T. and Hélène, C. (1968). *Nature,* **217,** 844.
20. Slifkin, M. A. Unpublished work.
21. Shinitzky, M., Katchalski, E., Grisaro, V. and Sharon, N. (1966). *Arch. Biochem. Biophys.* **116,** 332.
22. Shinitzky, M. and Katchalski, E. (1968). "Molecular Associations in Biology" (Ed. B. Pullman), p. 362. Academic Press, New York.
23. Smith, N. (1954). Ph.D. thesis, University of Chicago.
24. Shinitzky, M. and Goldman, R. (1967). *Europ. J. Chem.* **3,** 139.
25. Shinitzky, M. (1968). *Israel J. Chem.* **6,** 525.
26. Alivisatos, S. G. A., Ungar, F., Jibril, A. and Mourkides, G. A. (1961). *Biochim. biophys. Acta,* **51,** 361.
27. Cilento, G. and Guisti, P. (1959). *J. Am. chem. Soc.* **81,** 3801.
28. Cilento, G. and Tedeschi, P. (1964). *J. biol. Chem.* **236,** 907.
29. Shifrin, S. (1963). *Biochim. biophys. Acta,* **81,** 205.
30. Shifrin, S. (1964). *Biochemistry,* **3,** 829.
31. Foster, R. and Fyfe, C. A. (1966). *J. chem. Soc.* (B), 926.
32. Galzigna, L. (1969). *Biochem. Pharmacol.* **18,** 2485.
33. Galzigna, L. (1969). *FEBS Letters,* **3,** 97.

CHAPTER 6

Porphyrins

6.1. Introduction

The porphyrins are a very important class of biological molecule. Amongst the more important porphyrins are chlorophyll and heme, a chelate of iron and protoporphyrin, which occurs in the blood.

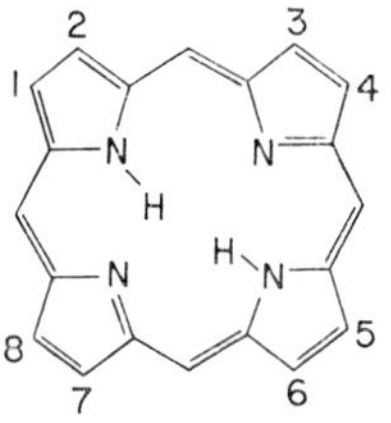

	1	2	3	4	5	6	7	8
hematoporphyrin	CH_3	$CHOH.CH_3$	CH_3	$CHOH.CH_3$	CH_3	C_2H_4COOH	C_2H_4COOH	CH_3
uroporphyrin	CH_3	$C_2H_3(COOH)_2$	CH_3	$C_2H_3(COOH)_2$	CH_3	$C_2H_3(COOH)_2$	$C_2H_3(COOH)_2$	CH_3
coproporphyrin	CH_3	C_2H_4COOH	CH_3	C_2H_4COOH	CH_3	C_2H_4COOH	C_2H_4COOH	CH_3
etioporphyrin	CH_3	C_2H_5	CH_3	C_2H_5	CH_3	C_2H_5	C_2H_5	CH_3

Fig. 6.1. Structure of some porphyrins.

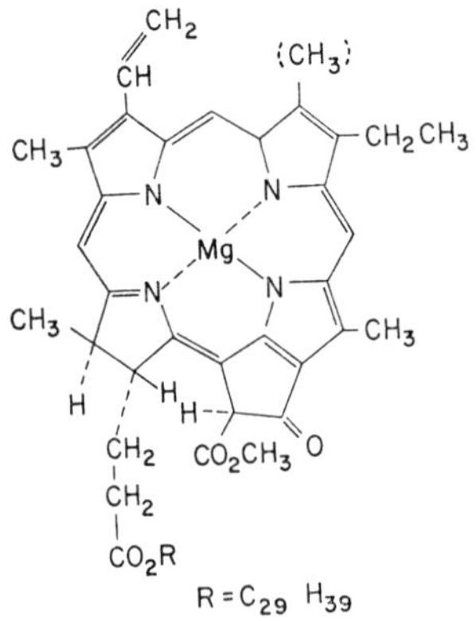

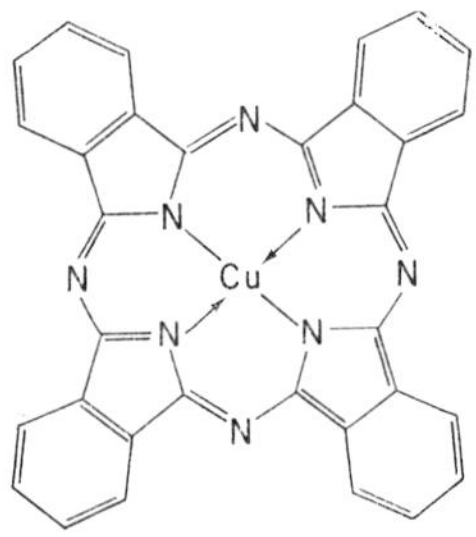

Fig. 6.1 *contd.* (A) Structure of chlorophyll a. For chlorophyll b the CH_3 group in the bracket is replaced by —COH. (B) Structure of copper phthalocyanine.

6.2. Chloranil

The earliest work to indicate that a porphyrin could act as a charge donor was done by Kearns, Tollin and Calvin (1). They showed that the addition of chloranil to the surface of films of metal-free phthalocyanine caused a very marked increase in the film's dark conductivity accompanied by a large ESR signal indicating the presence of free radicals. This ESR signal decreased on illumination but the conductivity increased. This is explained by the transfer of an electron from the phthalocyanine in the dark giving rise to the chloranil anion and the further transfer of an electron in the light giving the double anion. Such behaviour has been observed in organic charge transfer complexes. Similar effects which occur on adding oxygen to methylphthalo-cyanine probably arise from the same cause (2).

TABLE 6.1

Electric and magnetic properties of metal-free phthalocyanine-acceptor lamellae systems

1	2	3	4	5	6	7	8
Donor molecule solid state ionization potential, (eV)	Acceptor molecule	$g-$ Value $\pm$ 0·0002	Line width (gauss)	No. of unpaired electrons per donor molecule	Effect of illumination on the unpaired spin concn. (approx. % change)	Photo ESR decay constant, τ (sec) ($N = N_0$ $e^{-t/\tau}$)	Photo-current decay constant, τ (sec) ($N = N_0$ $e^{-t/\tau}$)
Metal-free phthalo-cyanine (4.5)	o-chloranil	2.0028	4.2 S[b]	0.002	D[c] (10%)	65	61
	iodine	2.0030	6.7 A[b]	0.01	D[c] (26%)	Nm[a]	Nm[a]

[a] Nm = no measurement.
[b] S = symmetric line shape, A = asymmetric line shape.
[c] D = decrease.
Reproduced with permission from Table 2, ref. 3.

Kearns and Calvin have also studied the electric and magnetic properties of chloranil and iodine phthalocyanine complexes in a laminated solid arrangement both in the dark and in the light (3). These studies confirm the occurrence of a charge transfer interaction between the two molecules, Table 6.1.

6.3. Metal Ions

Evidence that porphyrins can act as charge donors is also given by the flash photolysis studies of Linschitz and Pekkarinen (4). They showed that the triplet states of some porphyrins and anthracene were quenched by metal ions which they presumed came about by a charge transfer mechanism with the porphyrins as donors.

6.4 Trinitrobenzene

Gouterman and Stevenson (5) have examined the absorption spectra of solutions of some porphyrins and trinitrobenzene in nitrobenzene. Reversible changes in spectra have been observed and interpreted as arising from charge transfer complexing, although no charge transfer bands as such were seen. Association constants and enthalpies of dissociation have been evaluated. Of the three porphyrins studied, etioporphyrin formed the strongest complex,

TABLE 6.2

Complexes of trinitrobenzene with porphyrins

Donor	$\Delta F°$	$\Delta H°$	$T\Delta S°$	$K_c (M^{-1})$		ref.
etioporphyrin	− 2.9	− 6.5	− 3.6		a	5
tetraphenylporphyrin	− 2.3	− 5.3	− 3.0		a	5
Zn tetraphenylporphin	− 1.1			$0.6_{25°}$	a	5
hematoporphyrin		− 5.2			b	11
Co(II)mesoporphyrin dimethylester				120	c	12
Co(II)mesoporphyrin dimethylester				102	f	12
chlorophyll a	− 3.3	− 4.08	− 0.84		d	17
pheophytin a	− 2.6	− 4.47	− 1.84		d	17
chlorophyll b	− 2.34	− 4.45	− 2.08		d	17
pheophytin b	− 1.99	− 3.85	− 1.83		d	17
Zn phthalocyanine				10	e	6

a In nitrobenzene.
b In ethanol/toluene (20 : 80, v/v).
c In chloroform
d In diethyl ether.
e In acetone.
f In dichloromethane.
Thermodynamic parameters in Kcal/mole.

tetraphenylporphin the next strongest and Zn tetraphenylporphin the weakest. It is to be expected that etioporphyrin with eight electron donating alkyl groups on the pyrrole ring would be a stronger donor than tetraphenyl-

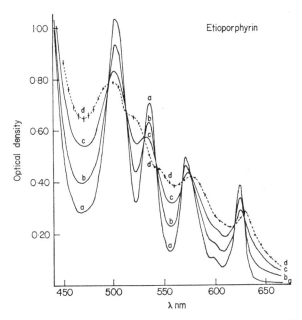

Fig. 6.2. Absorption spectra of etioporphyrin (concentration 7.0×10^{-5} moles/litre) in nitrobenzene solvent: (a) No TNB; (b) TNB, 0.04 moles/litre; (c) TNB, 0.20 moles/litre; (d) extrapolated pure complex absorption. Reproduced with permission from Fig. 1. ref. 5.

porphin. In a private communication, Gouterman has pointed out that these porphyrins agglomerate in solution and that there may be some error in the numerical results.

6.5. Aromatic Nitrocompounds

McCartin (6) has investigated the complexes formed between Zn phthalo-cyanine and various aromatic nitro compounds, using absorption spectro-scopy and fluorescence quenching. The absorption spectra of the complexes exhibited depression and broadening of the phthalocyanine bands similar to those observed for other porphyrin complexes (5). The fluorescence spectra of these complexes were quenched as compared to the free porphyrins, but no change in the spectra was detected. Association constants evaluated from absorption spectroscopy were different from those derived from fluorescence quenching and this difference attributed to the formation of excited state charge transfer complexes as well as ground state complexes.

6.6. Purines and Organic Cations

Mauzerall (7) has studied the changes in absorption and emission spectra of different porphyrins on forming molecular complexes in solution with various electron donors, namely the purines, adenine and caffeine and some large organic cations. The spectral changes are consistent with the formation of 1 : 1 complexes. Although Mauzerall did not interpret these complexes as charge transfer complexes, nevertheless in view of the porphyrins being

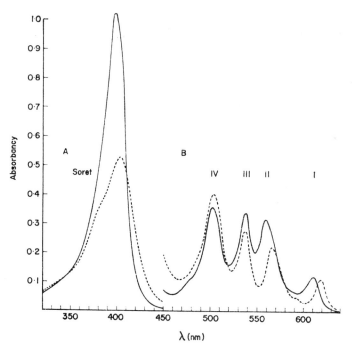

Fig. 6.3. Complex between uroporphyrin (URO) and 1-(carbamidomethyl)-3-carbamoyl-pyridinium chloride (CAMNACl). (A) Solid line: URO, 5×10^{-6} M in 0.1 M citrate, pH 5.9; dotted line: plus 0.037 M CAMNACl. (B) Solid line: URO, 3.5×10^{-5} M in 0.1 M EDTA, pH 6.0; dotted line: plus 0.095 M CAMNACl. Reproduced with permission from Fig. 1, ref. 7.

electron acceptors and because of the similarity with other complexes it seems likely that charge transfer is involved in these complexes. A much earlier worker Keilin (8) also found evidence for an interaction between caffeine and the porphyrins, although this was not interpreted as arising from charge transfer complexing.

6.7. Electron Donors and Acceptors

Cann (9, 10, 23) has investigated complexes formed between various electron donors and haemin, haematoporphyrin and myoglobin. Mixed solutions of these molecules in buffer, show spectral changes uncharacteristic of the individual molecules. In general, the changes are shifts in the porphyrin absorption bands as observed by other workers (6, 7), the exception being chlorpromazine (10), β-naphthoate and 3-indolebutyrate which also give

TABLE 6.3

Haematoporphyrin complexes. Comparison of association constants derived from quenching studies and absorption studies

Acceptor	$K_q (M^{-1})$	$K_c (M^{-1})$
trinitrotoluene	52	2.3
trinitrobenzene	75	10
trinitrofluorenone	very large	345
All measurements in acetone at 25°C		

Reproduced with permission from Table 1, ref. 6.

charge transfer bands in the red (23). In contrast, two electron acceptors nitrobenzene and iodonitrobenzene form 1 : 1 complexes with haematoporphyrin in ethanol, the stronger acceptor, iodonitrobenzene forming the stronger complex. The particularly strong electron donor tetramethylphenylenediamine also forms a 1 : 1 complex with haematoporphyrin. These

TABLE 6.4

Haematoporphyrin complexes

Compound	$\Delta H°$ (kcal/mole)	$K_c (M^{-1})$	solvent	ref.
nitrobenzene	—	$3.10^{-2}_{28°}$	a	10
iodobenzene	—	$11.10^{-2}_{28°}$	a	10
tryptophan	− 5.2	$7.18_{21°}$	b	11
6-aminocaproic acid	− 5.5	$1.13_{21°}$	b	11
arginine	− 5.0	$12.35_{24°}$	b	11
caffeine	− 10.8	$2220_{28°}$	b	11
6-methylaminopurine	− 3.5	$150_{27°}$	b	11
trinitrobenzene	− 5.2	$102_{21°}$	c	11
trinitrobenzene		$6.06_{23°}$	d	11
chloranil		$588_{23°}$	d	11

[a] Ethanol.
[b] pH 9 buffer.
[c] Ethanol/toluene (20 : 80, v/v).
[d] 50% ethanol.

various aromatic donors and acceptors all cause quenching of the porphyrin fluorescence, a characteristic of complex formation. It is clearly established that certainly haematoporphyrin and probably other porphyrins can act both as electron donors and acceptors under what may be described as moderate conditions. This is one of the few classes of biomolecules for which good experimental evidence exists for this ambivalent behaviour.

Experiments on the donor and acceptor behaviour of porphyrins have also been carried out by Heathcote, Slifkin and coworkers (11).

The absorption spectra of mixtures of haematoporphyrin and donors and acceptors have been studied. It was found that 1 : 1 complexes were formed with many, although not all donors and acceptors studied. With ionizable donors and acceptors, complexes were only formed in buffered solutions otherwise dissociation took place in preference to complex formation. Among the donors complexing with the porphyrin were various amino acids including tryptophan but excluding aliphatic α amino acids. Among the acceptors forming complexes were trinitrobenzene and chloranil. Enthalpies of dissociations were found to be consistent with values obtained by Gouterman and Stevenson (5) for other trinitrobenzene porphyrin complexes. An interesting point was the non-interaction of indole with haematoporphyrin which was taken to mean that tryptophan interacted as an n-donor via a lone pair electron on the nitrogen of the amino group rather than a π-electron donor from the indole ring system. Cann (23) however has shown that some indole derivatives not containing amino groups do complex with haemin.

Hill, MacFarlane and Williams (12) have isolated solid molecular complexes formed between some porphyrins and various aromatic nitro compounds. Elemental analysis established the 1 : 1 stoichiometry of these complexes. The association constants of some of these complexes have been measured.

6.8. NMR Studies

Hill, Mann and Williams (13) have investigated the effect of the same porphyrins on the NMR spectra of the aromatic nitro compounds. Signal shifts of the hydrogens in the aromatic nitro compounds are attributed to the formation of complexes and association constants derived from these shifts are similar to those derived spectrophotometrically.

These workers have continued their studies on the interaction of several metal porphyrins with different electron donors and acceptors including steroids (14). Although it was established that complexes were formed with both donors and acceptors, no correlation was found between electron affinities or ionization potentials and association constants. As a consequence the authors do not think that charge transfer forces are important in these complexes.

6.9. Vitamin B_{12} as an Acceptor and Vitamin B_{12b} as a Donor

Two molecules closely related to the porphyrins, namely vitamin B_{12} and vitamin B_{12b} have been studied by Heathcote and Slifkin (15) who found

TABLE 6.5

Complexes of Cobalt(II)mesoporphyrin dimethyl ester

Compound	$K_c\,(\text{M}^{-1})$	solvent	ref.
2,4,7-trinitrofluorenone	2500	a	12
trinitrofluorenone	970	b	14
trinitrofluorenone	3038	a	14
1,3,5-trinitrobenzene	120	a	12
trinitrobenzene	102	b	14
trinitrobenzene	160	a	14
3,5-dinitrobenzonitrile	95	a	12
dinitrobenzonitrile	40	b	14
dinitrobenzonitrile	104	a	12
2,4,6-trinitrotoluene	50	a	12
trinitrotoluene	10	b	14
trinitrotoluene	46	a	12
benztrifuroxan	670	b	14
dinitrobenzfuroxan	155	b	14
vitamin K_3	3.8	b	14
vitamin K_3	16	c	14
trinitrofluorenone	3000	c	14

^{a} chloroform.
^{b} dichloromethane.
^{c} ether.

TABLE 6.6

Complexes of Mn(III)porphyrin

Compound	$K_c\,(\text{M}^{-1})$		solvent
dimethylaniline	ca.	1	a
p-dimethoxybenzene		0.2	a
hexamethylbenzene		2.5	a
cholestane		3.1	a
cholestanone		40	a
cholestene		75	a
cholestanol		250	a
progesterone	ca.	4	a
trinitrofluorenone		250	a
trinitrobenzene		30	b

Reproduced with permission from Table 4, ref. 14.
^{a} In dichloromethane.
^{b} In chloroform.

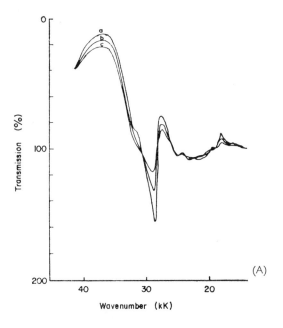

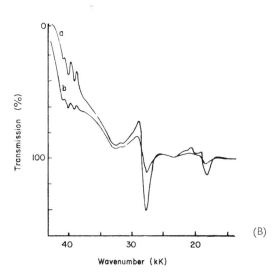

Fig. 6.4. (A) Charge transfer interactions. Difference spectra of $4.8 \cdot 10^{-3}$ M 8-aza-adenine and B_{12b} in 50% ethanol (a) at 25°, (b) at 32° and (c) at 45°. (B) Charge transfer interactions. Difference spectra of (a) $2.61 \cdot 10^{-1}$ M glutaric acid and B_{12} in 50% ethanol and (b) $8.7 \cdot 10^{-2}$ M glutaric acid and B_{12} in 50% ethanol. Reproduced with permission from Figs 1 and 3, ref. 15.

TABLE 6.7

Complexes of vitamin B_{12b}

Donor	$\Delta H°$ (kcal/mole)	K_c (M^{-1})	solvent	ref.
glycine	−1.52	76.8$_{25°}$	a	15
triethylamine	−7.64	141$_{25°}$	a	15
8-azaadenine	−26.8	214$_{25°}$	b	15
glycylglycine	−1.72	166$_{25°}$	a	c

[a] pH 7 buffer.
[b] 50% ethanol.
[c] Heathcote, J. G., Moxon, G. H. and Slifkin, M. A., (197). Spectrochim Acta, **27A**, 1391.

that vitamin B_{12} behaves as a charge donor in the presence of such charge acceptors as glutaric acid, alloxan, iodine and chloranil. Vitamin B_{12b} behaves as a charge acceptor in the presence of the amino acids (see Section 3.6), aliphatic amines, purines and pyrimidines. Enthalpies of dissociation have been evaluated for some of the complexes formed between these

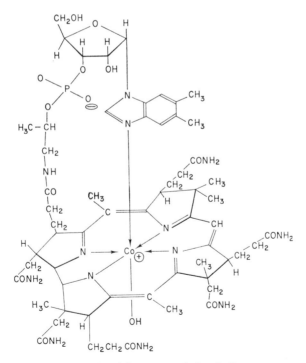

Fig. 6.5. Spatial structure of vitamin B_{12b}.

molecules. A purine 8-azaadenine forms a remarkably strong complex with vitamin B_{12b} having the very high value of enthalpy of dissociation of -27 kcal/mole. These results are contrary to the predictions of Veillard and Pullman (16) who suggested that vitamin B_{12} should be an electron acceptor.

An interesting feature of these complexes is that the spectrum of the vitamin B_{12} complex is identical to that of vitamin B_{12b} and the spectrum of the vitamin B_{12b} complex is identical to that of vitamin B_{12}. This can be explained as the vitamin B_{12b} contains the cobalamin cation and vitamin B_{12b} neutral cobalamin represented thus

spectrum of B_{12} electron donor or   spectrum of B_{12b}
negative ion
$\rightleftharpoons$
$Cob(H_3O)^+$ electron acceptor or   CobCN
removal of negative ion

Complexes of dipeptides with the vitamin are discussed in Section 3.6 as are the infra-red spectra and mechanisms of interaction.

6.10. Chlorophyll

6.10.1 CHLOROPHYLL AND TRINITROBENZENE

The charge transfer properties of chlorophylls have been studied by various workers. Larry and Van Winkle (17) have looked at the interaction of some chlorophylls and pheophytins with the well-known acceptor trinitrobenzene. Association constants and thermodynamic parameters have been evaluated from visible and ultra-violet absorption spectroscopy. Although no charge transfer bands were seen, only the typical broadening and decrease of absorption typical of porphyrin complexes, these complexes were taken to be charge transfer complexes. The infra-red spectra of these complexes showed small changes in the bands arising from the $C=O$ groups which were shifted by small amounts *ca.* 15 cm^{-1}. NMR studies of these complexes showed shifts of the lines observed in free chlorophyll. The maximum amount of shift occurs when trinitrobenzene and chlorophyll are present in equimolar proportion confirming the 1 : 1 stoichiometry of the complexes. The electron donating ability of the molecules is chlorophyll a > pheophytin a > chlorophyll b > pheophytin b (see Table 6.2).

6.10.2. ESR OF CHLOROPHYLL COMPLEXES

Tolli and Green have studied the effect of light on the ESR signals of mixtures of chlorophyll a and quinones (18). Only in the presence of light is a large signal seen and this arises from the semiquinone. The action spectrum for this signal is quite similar to the absorption spectrum of the chlorophyll. It would appear therefore that the excited chlorophyll is functioning as an electron donor. The authors suggest that in the case of o-chloranil which has a particularly wide ESR signal in the light that a complex is formed between the chlorophyll cation and the semiquinone. A similar explanation is advanced for the complex with a biological quinone coenzyme Q_6. This is shown schematically;

$$C + h\nu = C^*, \; C^* + \text{quinone} \rightarrow (C^+ \ldots \text{semiquinone})^* =$$
$$= C^+ + \text{semiquinone} \rightleftharpoons C + \text{quinone}.$$

the bracketed term represents a complex formed between the excited chlorophyll C^* and the semiquinone with vibrational energy of excitation.

At low temperature it is suggested that a ground state complex is formed between the chlorophyll and quinone before the formation of the excited state complex:

$$C + \text{quinone} \rightleftharpoons (C \ldots \text{quinone})$$
$$(C \ldots \text{quinone}) + h\nu = (C^+ \ldots \text{semiquinone})^*.$$

Later work (19) on the effect of adding other acceptors to these systems showed that the semiquinone of the most electron negative acceptor in the system were produced on illumination. NADH caused an eight-fold enhancement of the free radical concentration. This is explained by the regeneration of neutral chlorophyll from its ion by the oxidation of NADH, the oxidized NADH then reacting with the acceptor anion to produce NADH and neutral acceptor. The action of light on mixtures of chlorophyll a and a biological quinone riboflavin is to produce the riboflavin semiquinone as detected by ESR spectroscopy and ultra-violet spectrophotometry. However, there is no evidence of complex formation in these systems.

6.10.3. ELECTRICAL CONDUCTIVITY OF CHLOROPHYLL COMPLEXES

Electrical conductivity measurements have been carried out on mixtures of chlorophyll with various biomolecules, in the solid state (20). Changes in conductivity of the mixtures compared to the separate components suggest that definite complex formation occurs. Whereas chlorophyll has a resistivity of 3.10^{12} ohm cm, when complexed with bovine plasma albumin the resistivity drops to 5.10^{11} ohm cm, bovine plasma albumin itself having a resistivity of 5.10^{17} ohm cm. The respective activation energies are 1.45 eV, 0.51 eV and 2.86 eV. Complexes also occur with β-carotene and β-methylnaphthaquinone.

6.11. Conductimetric Titration of Iodine Complexes

Conductimetric titration of phthalocyanine (metal-free) with iodine in dimethyl sulphoxide has established the formation of 1 : 2 complexes between phthalocyanine and iodine (21). This is unexpected as the stoichiometry of all the other porphyrins would appear to be 1 : 1.

6.12. Inhibition of Urea Denaturation

Cann has investigated the effect of aromatic compounds, i.e. π-donors on the urea denaturation by myoglobin, a porphyrin-containing protein (9), and on the zinc-myoglobin interaction (22,23). These aromatic compounds enhance the denaturation as determined by the change in absorbance of the protein in the very near ultra-violet region, i.e. the Soret band (9). The effect of these compounds correlates well with their ionization potentials. It is therefore concluded that these compounds form charge transfer complexes with some part of myoglobin.

The rate of enhancement of the aromatic donors in the zinc-myoglobin interaction has been followed spectrophotometrically and this enhancement correlates with decreasing ionization potential of the donors (21). It is proposed that the formation of charge transfer complexes relieves intramolecular π-bonding interactions in the protein at specific sites hence decreasing the activation required for the zinc-myoglobin interaction.

6.13. Effect of Electron Accepting Gases on the Semiconductivity of Porphyrins

Several authors have shown that the semiconductivity of porphyrins is greatly affected by the presence of absorbed gases such as oxygen and nitric oxide which are electron acceptors (24, 25, 26).

The semiconductivity of chlorophyll a and b is markedly changed by surface absorbed oxygen (24). Oxygen causes a rise both in dark current and photocurrent ascribed to the formation of a complex. The binding energy of the complexes are for chlorophyll a, 1.4 eV and chlorophyll b 0.63 eV. There is no effect on the activation energy of the semiconductivity. This behaviour is unlike that of other conventional charge transfer complexes and the binding energies are very much migher than those normally associated with charge transfer complexes.

Conversely Kaufhold and Hauffe (25) show that the effect of electron accepting gases on thin films of phthalocyanines is to lower both the resistivity and the activation energy. A contrary result has been noted by Epstein and Wildi who report that the removal of oxygen from polymeric copper phthalocyanine causes a sharp drop in the resistivity with a concomitant drop of the activation energy (26). The effect of oxygen is very unclear and it is difficult to draw any conclusions about the mode of interaction except to say that it is a function of its electron accepting property.

6.14. Conclusion

The porphyrins behave both as electron donors and acceptors under moderate conditions. This is one of the few classes of molecules which show this behaviour.

REFERENCES

1. Kearns, D. R., Tollin, G. and Calvin, M. (1960). *J. chem. Phys.* **32**, 1020.
2. Vartanyan, A. T. and Karpovich, I. A. (1956). *Dok. Akad. S.S.R.* **111**, 561.
3. Kearns, D. R. and Calvin, M. (1961). *J. Am. chem. Soc.* **83**, 2110.
4. Linschitz, H. and Pekkarinen, L. (1960). *J. Am. chem. Soc.* **82**, 2411.
5. Gouterman, M. and Stevenson, P. E. (1962). *J. chem. Phys.* **37**, 2266.
6. Cartin, P. J. (1963). *J. Am. chem. Soc.* **85**, 2021.
7. Mauzerall, D. (1965). *Biochemistry*, **4**, 1801.
8. Keilin, J. (1943). *Biochem. J.* **37**, 281.
9. Cann, J. R. (1967). *Biochemistry*, **6**, 3427.
10. Cann, J. R. (1967). *Biochemistry*, **6**, 3435.
11. Heathcote, J. G., Hill, G. J., Rothwell, P. and Slifkin, M. A. (1968). *Biochim. biophys. Acta*, **153**, 13.
12. Hill, H. A. O., MacFarlane, A. J. and Williams, R. J. P. (1967). *Chem. Comm.* 905.
13. Hill, H. A. O., Mann, B. E. and Williams, R. J. P. (1967). *Chem. Comm.* 906.
14. Hill, H. A. O., MacFarlane, A. J. and Williams, R. J. P. (1969). *J. chem. Soc.* (A), 1704.
15. Heathcote, J. G. and Slifkin, M. A. (1968). *Biochim. biophys. Acta*, **158**, 167.
16. Veillard, A. and Pullman, B. (1965). *J. Theoret. Biol.* **8**, 307.
17. Larry, J. R. and van Winkle, Q. (1969). *J. phys. Chem.* **73**, 570.
18. Tollin, G. and Green, G. (1962). *Biochim. biophys. Acta*, **60**, 524.
19. Tollin, G. and Green, G. (1963). *Biochim. biophys. Acta*, **66**, 308.
20. Eley, D. D. and Snart, R. S. (1965). *Biochim. Biophys. Acta*, **102**, 379.
21. Guttmann, F. and Keyzer, H. (1966). *Electrochim. Acta*, **11**, 555.
22. Cann, J. R. (1965). *Biochemistry*, **4**, 2368.
23. Cann, J. R. (1969). *Biochemistry*, **8**, 4036.
24. Rosenberg, B. and Camiscoli, J. F. (1961). *J. chem. Phys.* **35**, 982.
25. Kaufhold, J. and Hauffe, K. (1965). *Ber. Bunseng. Phys. Chem.* **69**, 168.
26. Epstein, A. and Wildi, B. S. (1960). *J. chem. Phys.* **32**, 324.

CHAPTER 7

The Flavins

7.1 Introduction

The flavins are probably the most extensively studied of the biomolecules able to form complexes. This reflects the biological importance of this group. Riboflavin (RFN) is important nutritionally as the vitamin B_2. Flavins such as flavin mononucleotide (FMN) and flavin adenine dinucleotide (FAD) are present in the mitochondria and take part in oxidative phosphorylation.

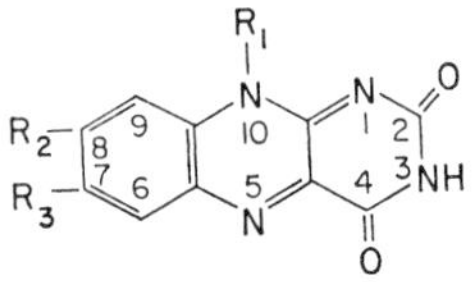

R_1	R_2	R_3	
H	H	H	isoalloxazine
H	CH_3	CH_3	lumichrome
CH_3	CH_3	CH_3	lumiflavin
ribityl	CH_3	CH_3	riboflavin
ribityl-5′-phosphate	CH_3	CH_3	flavin mononucleotide
ribityl-5′-adenosine monophosphate	CH_3	CH_3	flavin adenine dinucleotide

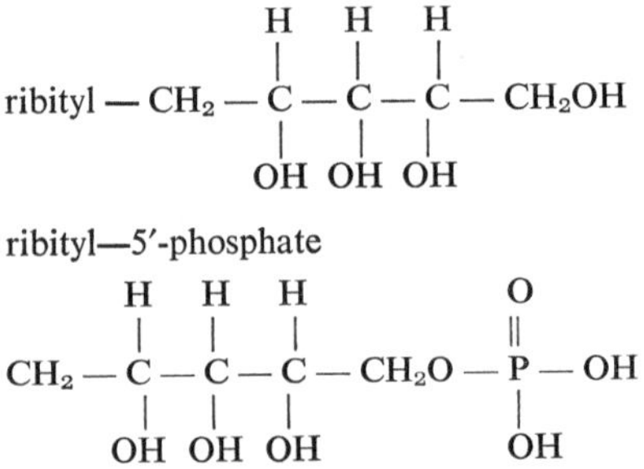

Fig. 7.1. Structures of some flavins.

The flavins are derivatives of isoalloxazine and are thus quinones. In view of the well-established electron acceptor character of organic quinones such as chloranil, they would certainly be expected to function as electron acceptors.

Flavins can exist in various states as they undergo protonation, oxidation and reduction in solution. In effect there are nine possible forms. These consist of the normal fully oxidized form which can also exist in a monoprotonated and a deprotonated form, depending on the pH. There are similarly three fully reduced forms and an intermediate form between the fully oxidized and fully reduced form, the semiquinone.

$$FlH_2^+ \underset{pK_a \sim 0}{\rightleftharpoons} FlH \underset{pK_a \sim 10}{\rightleftharpoons} Fl^- \qquad \text{quinone}$$

$$FlH_3^+ \underset{.pK_a \sim 1-3}{\rightleftharpoons} FlH_2 \underset{.pK_a \sim 6.5}{\rightleftharpoons} FlH^{-\cdot} \qquad \text{semiquinone}$$

$$FlH_4^+ \underset{pK_a < 0}{\rightleftharpoons} FlH_3 \underset{pK_a \sim 6.2}{\rightleftharpoons} FlH_2^- \qquad \text{hydroquinone.}$$

The interactions of the flavins will be discussed with different groups of molecules separately.

7.2 Indoles

It was observed a long time ago that RFN could be solubilized by tryptophan in water (1). This is quoted in the Pharmacopoeia (2) as a method of preparing RFN solutions for pharmacological purposes.

Harbury and Foley (3) measured the apparent association constants of several compounds including tryptophan with FMN and 3-methyl RFN. These constants are of the order of 20 M^{-1}. The constants were evaluated using a spectrophotometric method as the addition of the great number of molecules to be discussed in this chapter is to cause the same changes in the spectrum of the flavins. There is a decrease in the absorption of the main flavin band at *ca.* 470 nm with a slight increase in absorption on the long wavelength side so that there is tailing off into the red at about 600 nm (Fig. 7.2). As both the flavins studied by Harbury and Foley have the same association constant, it is presumed that the force binding the complex cannot be due to hydrogen bonding, methylation of RFN removes the available proton at the 3 position, and probably arises from the formation of 1 : 1 charge transfer complexes.

Isenberg and Szent-Györgyi (4) have carried out similar experiments on the complexes formed between tryptophan and serotonin with FMN. The association constants for the respective complexes are 60 M^{-1} and 400 M^{-1}. Freezing these complexes produces a very marked intensification of colour which disappears on rewarming. Similar colours can be produced by preparing solid complexes. The colour plate (Fig. 7.3) shows some complexes which were made by evaporating equimolar solution of donor and flavin to

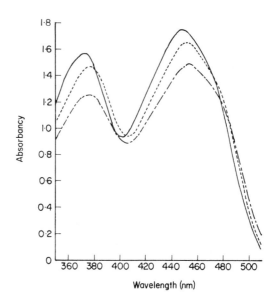

Fig. 7.2. Absorption spectra of solutions of 3-methylriboflavin plus interactant. Cary model 11 spectrophotometer, 10 cm. cuvettes. ————, 1.5 × 10⁻⁵ M 3-methylriboflavin *vs.* water; ————, 1.5 × 10⁻⁵ M 3-methylriboflavin plus 0.21 M benzoate *vs.* 0.21 M benzoate; —–—, 1.5 × 10⁻⁵ M 3-methylriboflavin plus 0.020 M 2-naphthoate *vs.* 0.020 M 2-naphthoate. Reproduced with permission from Fig. 2, ref. 3.

dryness at low pressure and temperature in a rotary evaporator (5). The infra-red spectrum of a solid tryptophan RFN complex prepared in this manner shows a slight shift of the carbonyl bands to longer wavelengths (5).

Crystalline complexes of indoles and flavins prepared by Pereira and Tollin are also highly coloured (6). The spectra of these crystalline complexes in mineral oil mulls show that the colour originates from absorption bands in the region of 550 nm which are specific to each complex (Figs. 7.4 and 7.5).

Detailed studies of the association constants of indole flavin complexes in different solvents at different pH have been reported several times.

Pereira and Tollin (6) have recorded the association constants and thermodynamic parameters for complexes in both neutral solution and acid solution (Table 7.3).

Wilson (7) has presented thermodynamic parameters for some indole FMN complexes and has pointed out that the indoles do not affect the spectrum of fully reduced flavin.

Changes in absorbance in the spectrum of RFN on adding indoles among other biomolecules are listed by Wright and McCormick (8).

From the above mentioned studies some interesting facts emerge. The

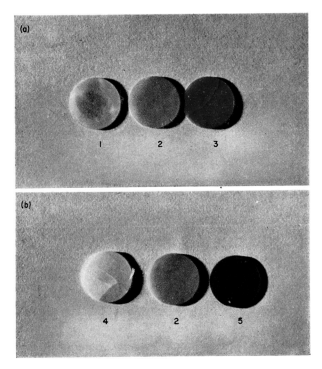

Fig. 7.3. (a) 200 mgm KBr discs containing: (1) 0.5 mgm tryptophan, (2) 0.5 mgm ribo-flavin, (3) 1.0 mgm tryptophan-riboflavin complex. (b) 200 mgm KBr discs containing: (4) 0.5 mgm caffeine, (2) 0.5 mgm riboflavin, (5) 1.0 mgm caffeine-riboflavin complex.

[facing p. 134

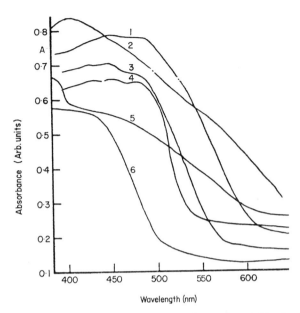

Fig. 7.4. Absorption spectra at room temperature of several crystalline lumiflavin–indole complexes dispersed in mineral oil. Baselines have been shifted for clarity. 1, Lumiflavin–3-methylindole (neutral); 2, lumiflavin–tryptophan (acid); 3, lumiflavin–indole (neutral); 4, lumiflavin alone (neutral); 5, lumiflavin–carbazole (acid); 6, 9-methylisoalloxazine hydrochloride. Reproduced with permission from Fig. 1, ref. 6.

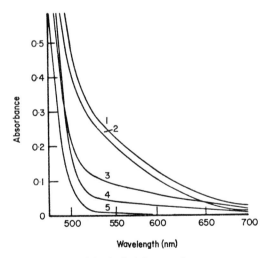

Fig. 7.5. Absorption spectra of flavin–indole complexes at room temperature in acid solution. 1, Riboflavin–tryptophan; 2, lumiflavin–tryptophan; 3, lumiflavin–1,2-dimethyl-indole; 4, riboflavin; 5, lumiflavin. Reproduced with permission from Fig. 2, ref. 6.

stability of flavin complexes with a given indole has the order FMN = lumi-flavin > RFN in acid and the stability of indoles with a given flavin is 2-methylindole > 1,2-dimethylindole > 3-methylindole (6). In neutral solution however, the order is 2-methylindole > 3-methylindole > 1,2-dimethylindole (6). This is different to the order of complexing found in chloranil complexes by Foster and Hanson (Ref. 5, Ch. 5), which is 1,2-dimethylindole > 2-methylindole > 3-methylindole, the association constants being taken as the measure of stability. These flavin complexes all possess 1 : 1 stoichiometry and are dissociated to some extent on the addition of ethanol. The addition of ethanol to water of course lowers the dielectric constant of the solvent. There does not seem to be any structural requirements for complexing as can be seen by examining the parameters of the different indole complexes. There is no correlation between the ionization potentials of the indoles and the position of maximum absorption of the new bands of the complexes.

ESR measurements by Isenberg, Szent-Györgyi and Baird (9) on the serotonin FMN complex in acid solution has been interpreted as being due to the flavin semiquinone. However Pereira and Tollin (6) claim that these signals are in fact light activated and that the complexing actually inhibits the ESR signal rather than produces it, by quenching the flavin triplet state. No signals were observed from the serotonin FMN complex in neutral solution.

Intra-molecular interactions between tryptophyl residues and flavins have been observed in synthetic flavinylpeptides (10, 11). The spectra of these peptides are very similar to mixtures of free flavins and indoles. The flavin fluorescence is quenched as compared to the fluorescence of the free flavin, the degree of quenching being a function of the solvent. The order of quenching is water > ethanol > chloroform > dimethylformamide. The observation of fluorescence quenching as a function of chain length of the flavinyl-peptides, leads to the conclusion that two types of quenching mechanisms are in operation; one due to the formation of ground state complexes and the other due to the formation of excited state complexes.

The complexing of FMN with tryptophyl residues in peptides has been utilized by Swinehart and Hess (12) to distinguish between buried and exposed residues. Association constants of various FMN α chrymotrypsin complexes have been measured and all have a value of *ca.* 400 M^{-1}, a similar value to that of the FMN serotonin complex (4).

A quite different type of study has been carried out by Radda (13) who has observed the inhibition of the photoreduction of flavins by EDTA and NADH in the presence of indoles, purines and phenols among other compounds. In general those molecules which form complexes with FMN greatly inhibit the photoreduction, the effect decreasing with increasing temperature.

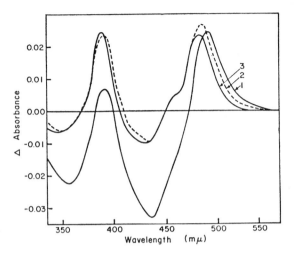

Fig. 7.6. Difference spectra of flavinyl peptides [$n = 1$] in 5×10^{-2} M sodium phosphate, pH 7.0. Notations are: 1, tryptophan; 2, tyrosine; 3, phenylalanine. Sample cuvette: 2×10^{-5} M flavinyl peptide; reference cuvette: 2×10^{-5} M ω-carboxyalkylflavin. Reproduced with permission from Fig. 1, ref. 10.

Conversely those molecules which do not complex with FMN have no inhibitory effect. It is suggested that the inhibition occurs by the formation of an excited state complex between the excited triplet state of the flavin and the inhibitor, as the inhibitors have no effect on dark photoreduction (see p. 136). The mechanism is represented schematically thus;

$$\text{donor}^{3*} + \text{acceptor}^1 \rightarrow \text{complex}^{3*} \rightarrow \text{complex}^1 \rightarrow \text{donor}^1 - \text{acceptor}^1,$$

the donor being the inhibitor, the flavin the acceptor. The dissociation of the excited triplet complex directly into the ground state donor and acceptor is thought to be the more likely mechanism as the triplet energy of FMN is much lower than the triplet energies of the majority of the inhibitors.

Photobleaching of FMN is also inhibited by the same indoles, phenols and purines. A similar mechanism to that just outlined presumably operates (14).

The effect of the addition of a variety of substances on the FMN NADH dark interaction has been reported by Fox and Tollin (15). These substances which include purines and indoles cause a decrease in the interaction rate which correlates with the ability of the added material to form complexes with the flavin. There is no evidence to suggest that FMN and NADH complex together in the dark.

Tryptophan quenches the triplet state of FMN produced during flash photolysis in 6 N HCl at 77°K by complexing (16).

7.3 Phenols

Phenols readily form complexes with the flavins. There does not appear to be any marked differences between phenols and indoles in interacting with flavins.

Harbury and Foley (3) have demonstrated that the absorption spectra and fluorescence of FMN, RFN and 3-methyl RFN is modified by phenols in just the same manner as tryptophan. The association constants of some phenol complexes are somewhat lower than tryptophan complexes. Similar results have also been found for the phenol RFN complex which has an association constant of 47 M^{-1} (17). More extended recent studies of the interaction of phenol derivatives with FAD concur with the earlier studies (18) (Tables 7.7 and 7.9).

Tollin and his coworkers have also carried many studies of phenol flavin complexes. The addition of electron-rich phenols to flavins in strongly acidic solution (12 N HCl) gives rise to very intense colours (19). The spectra of these mixed solution being again very similar to those of other flavin complexes. Spectra of solid crystalline complexes are similar to solution spectra but show an additional band in the red, the maximum being specific to each phenol derivative. Benzene derivatives are quite inactive so it can be assumed that the phenolic hydroxyl group is responsible for complexing. In neutral solution the spectra of the mixtures are the same as those displayed by the indole flavin mixtures. Isolated solid complexes, from neutral solution, also exhibit new bands in the long wavelength region whose positions are specific to the phenol derivative. The difference between complexes in neutral solution and in acidic solution is that the flavin is present in the monoprotonated form FlH_2^+ in the acid medium as shown by pH dependence studies of complexing. The neutral complexes have 1 : 1 stoichiometry in solution but 1 : 2 in the solid. The complexes obtained from the acidic solution have 1 : 1; 1 stoichiometry as the complex in fact has the structure phenol-flavin-HCl.

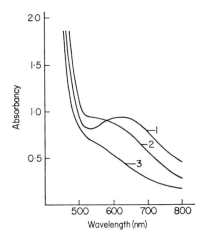

Fig. 7.7. Optical absorption of solutions of 9-methylisoalloxazine with a variety of phenols. Solvent was 50% 12 N HCl–50% ethanol. Flavin concentration was about 0.1 M, phenol about 0.2 M. 1, 1,4-Naphthalenediol; 2, 1,2-naphthalenediol; 3, trimethylhydroquinone. Reproduced with permission from Fig. 2, ref. 19.

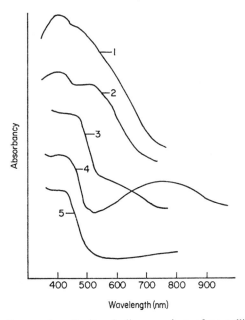

Fig. 7.8. Absorption spectra of mineral oil suspensions of crystalline acid complexes. 1, Riboflavin–hydroquinone; 2, 9-methylisoalloxazine–4-chlorocatechol; 3, 9-methyliso-alloxazine–1,2-naphthalenediol; 4, 9-methylisoalloxazine–1,4-naphthalenediol; 5, 9-methyl-isoalloxazine hydrochloride. Baselines have been shifted for clarity. Reproduced with permission from Fig. 3, ref. 19.

From the measurement of a number of association constants of phenol flavin complexes it is found that there is an inverse correlation between the association constants and the position of the new bands in the red (20). There is also a direct correlation between the energy of the highest filled molecular orbital of the phenols as determined theoretically and the position of the new absorption bands. The complexes as noted by other workers are more dissociated in less polar media. This dissociation is accompanied by a blue-shift of the band associated with the complex. The association constants are larger in neutral solution than in alkali solution. The order of stability of phenol complexes is FMN > lumiflavin, the opposite order to that found for indole complexes. (See p. 136, and Tables 7.1 and 7.2).

3,5-Diiodo-4-hydroxybenzoic acid binegative ion forms a complex with normal oxidized RFN and with the anionic form of RFN the complexes having association constants of 42.8 and 10.5 M^{-1} respectively (21). The corresponding unionized phenol, 3,5-diiodotyrosine has an association constant for its complex with normal RFN of 50 M^{-1}. The association constant of the complex with RFN in the binegative ionic form is 41.3 M^{-1}. Thus the dissociation of the diiodiphenols is unaffected by the formation of complexes with RFN.

Spectrophotometric data and thermodynamic parameters have been evaluated for the complexes formed between halogenated tyrosines and either FMN or RFN (22). Progressive halogenation causes an increase in the association constants. The changes in the enthalpies of dissociation are somewhat equivocable. Cilento and Berenholc assuming that the increasing in association constants implies increase in bonding energy, interpret this affect as due to back charge transfer with the flavins as donors and the tyrosines as acceptors. Another possible explanation is that the tyrosines act as n-donors rather than π-donors and that therefore the halogenation of the phenol ring does not have the effect of decreasing the donor ability of the tyrosine. (See Section 3.3.) The association constants are lower in alkali solution than acidic solution. Another indication of complexing in these systems is the quenching of the flavin fluorescence.

Riboflavin reacts with hydriodic acid, HI, to form a complex, the flavin being present as the semiquinone $FIH_3^{+\cdot}$ as shown by the absorption spectrum which has a well-defined peak at 500 nm (23). More complexes can be made on the addition of other molecules to solutions of RFN and HI, such as phenols and hydroquinone in which all three components are present in 1 : 1 : 1 stoichiometry. Crystals of these complexes exhibit strong para-magnetism with approximately 100% unpaired spin.

The electrical and magnetic properties of large single crystals of phenol-protonated flavin complexes have been studied together with a RFN-hydroquinone complex (24). The complexes all have a much lower specific

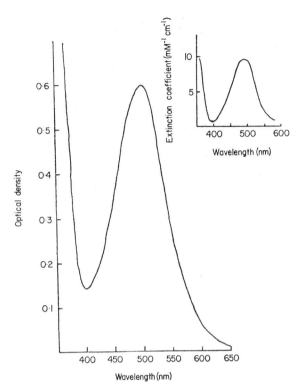

Fig. 7.9. Visible spectrum of a solution of riboflavin in 47% hydriodic acid. *Inset:* Absorption spectrum of FMN semiquinone at pH 0.4. Reproduced with permission from Fig. 1, ref. 23.

electrical resistivity than RFN alone. The activation energy of the RFN-hydroquinone complex is 1.1 eV, a somewhat higher value than those more common for organic complexes of hydrocarbons and quinones. Whilst these complexes all exhibit ESR signals these may arise from light activation (see p. 136).

Photobleaching of FMN is inhibited by phenols as is the photoreduction of flavins by EDTA and NADH (13, 14). The suggested mechanisms are as described on p. 137.

The crystal structures of two different complexes involving phenol-like molecules have been worked out by X-ray crystallography.

The riboflavin hydrogen bromide hydroquinone complex exists in two allotropic forms (61). One allotropic form has the hydroquinone overlapping the 8, 9, 10 positions of the isoalloxazine ring with a separation of 3.35 Å. In the other allotropic form there is a separation between the quinol and the flavin of 3.28 Å and the hydroquinone overlaps the 2, 3 positions of the isoalloxazine ring. In both forms there is overlap with electron deficient regions. Electron densities of these molecules have been calculated by the Pullmans (62). The separations of the allotropic forms agree with the criterion laid down in Section 1.4 for charge transfer complexes.

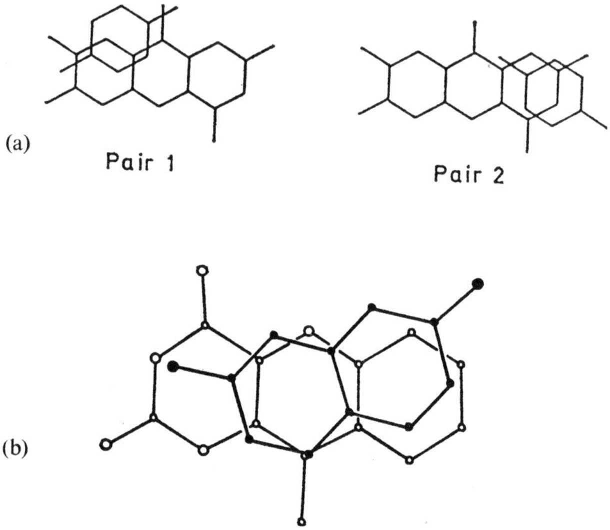

(a)

Pair 1   Pair 2

(b)

Fig. 7.10. Structures of flavin complexes. (a) Structures of two allotropic forms of the complex formed between riboflavin dihydrobromide and hydroquinone. Reproduced with permission from ref. 61. (b) Structure of the complex formed between 10-methylisoalloxazine bromide sesqui-naphthalene-2,7-diol monohydrate. Reproduced with permission from ref. 63.

The structure of the molecular complex formed between 10-methyliso-alloxazine bromide and sesqui-naphthalene-2,7-diol monohydrate has a rather different structure from the hydroquinone complex (63). Overlap occurs mainly in the N5, C8, N10, C6 region of the isoalloxazine ring, a region of low electron density.

7.4 Purines and Pyrimidines

The first direct evidence for the formation of complexes between purines and a flavin was given by the now classic work of Weber (25) who showed that the purines, caffeine and adenosine, quench the fluorescence of RFN. These quenching studies show that the complexes have 1 : 1 stoichiometry and are temperature reversible. Enthalpies of dissociation of both complexes are about −1.6 kcal/mole. The spectra of the complexes bear a close similarity to those reported for phenol and indole complexes with the flavins and to the spectrum of FAD. Furthermore, the fluorescence of FAD and its dependence on pH can be correlated with the fluorescence and pH dependence of complexes of FMN with adenine or adenosine. This leads to the inescapable conclusion that an intramolecular complex exists in FAD between the flavin moiety and the adenosine moiety in the molecule.

Burton (26) has measured the association constants of complexes formed between FAD and some purines using a similar fluorescence technique to Weber (see Table 7.8).

Harbury and Foley (3) have shown that purines will complex with 3-methyl RFN as well as with FMN and RFN.

The writer (27) has demonstrated that by no means all purines complex with RFN. Two notable exceptions as indicated from absorption spectroscopy are uric acid and 8-azaadenine, two normally reactive compounds. The spectral changes observed on adding the purines to solutions of RFN are a new positive peak appearing at about 300 nm with an isosbestic point at 330 nm. Negative peaks appear in the difference spectra at the same positions of maximum absorption of free RFN. In pH 4 buffer, the free RFN gradually reduces to the fully reduced form. The addition of the purines partially

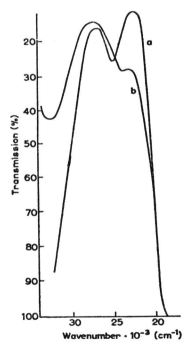

Fig. 7.11. Absorption spectra. Riboflavin in aqueous buffer (pH 4): (a) on mixing, (b) after 12 h. Reproduced with permission from Fig. 5, ref. 27.

stabilizes RFN against reduction so that the difference spectra now show peaks at the maximum position of the free RFN absorption, the same isosbestic point at 330 nm with a decrease in absorption to shorter wavelength.

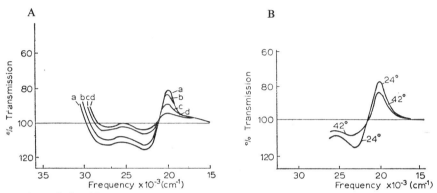

Fig. 7.12. A. Difference spectra. (a) $1.275.10^{-2}$ M tetramethyluric acid, (b) $7.72.10^{-3}$ M tetramethyluric acid, (c) $6.45.10^{-3}$ M tetramethyluric acid, (d) $3.96.10^{-3}$ M tetramethyluric acid, all in $1.15.10^{-4}$ M RFN *vs.* RFN in water. B: Difference spectra. $1.275.10^{-2}$ M tetramethyluric acid at 24°C, and $1.275.10^{-2}$ M tetramethyluric acid at 42°C, both in $1.15.10^{-4}$ M RFN *vs.* RFN in water. Reproduced with permission from Figs. 1 and 2, ref. 28.

The tetramethyluric acid RFN complex has been studied by the writer in greater detail and has an enthalpy of dissociation of -2.2 kcal/mole (28).

Wright and McCormick (8) have listed the decrease in absorbance of various molecules including caffeine and tryptophan on adding RFN. They have shown that complexing causes a marked change in the spectrum of RFN. The peak absorption of RFN at 266 nm is markedly diminished and the peak shifts to 280 nm.

A study has been made on the interaction of a large number of nucleosides and free bases with RFN to elucidate what are the structural requirements for complex formation (29). Using the fluorescence quenching technique it has been shown that substitution into the N9 position has a marked effect of the complex stability. Adenine is a stronger complexer with RFN than is adenosine which in turn is a stronger complexer than adenosine monophosphate. Similarly the behaviour of a number of free bases in complex forming is determined by substitution in the N9 position. In addition to this dependence there is a good correlation coefficient of correlation $= -0.87$, between the energy of the highest filled molecular orbital as determined theoretically and the stability of the complex, i.e. the association constants of the complexes as determined from quenching measurements (Table 7.1). Substitution into the 6-position of the purine or pyrimidine also increases the ability to complex with RFN as does increased alkylation of an amino group

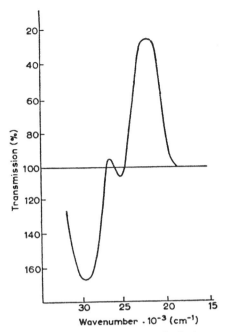

Fig. 7.13. Difference spectrum. Old adenine and riboflavin *vs.* riboflavin in aqueous buffer (pH 4). Reproduced with permission from Fig. 4, ref. 27.

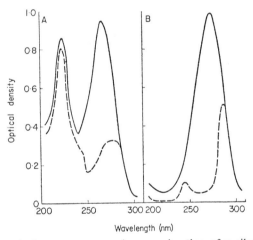

Fig. 7.14. Spectral changes accompanying complexation of D-riboflavin with 1,1,3-tricyano-2-amino-1-propene in 0.1 M sodium phosphate buffer, pH 7. In part A, the spectrum (solid line) of 30 μM riboflavin in buffer against a buffer blank is compared with the difference spectrum (dashed line) of 30 μM riboflavin plus 430 μM trinitrile in buffer against a buffer plus trinitrile blank. In part B, the spectrum (solid line) of 64 μM trinitrile in buffer against a buffer plus trinitrile blank. In part B, the spectrum (solid line) of 64μM trinitrile in buffer against a buffer blank is compared with the difference spectrum (dashed line) of 64 μM trinitrile plus 260 μM riboflavin in buffer against a buffer plus riboflavin blank. Reproduced with permission from Fig. 1, ref. 8.

TABLE 7.1

Data for riboflavin complexes in aqueous solution

Donor	pH	$T(°C)$	$K_c(M^{-1})$	$\Delta H°$ (kcal/mole)	Ref.
caffeine	alkali	20	41.8	− 5.74	33
caffeine	7	17	91	− 1.6	25
caffeine		20	52.6		17
tetramethyluric acid	5	24	33.6	− 2.2	28
2-methylindole	acidEt.	27	603	− 1.02	6
3-methylindole	acidEt.	27	35.5	− 1.92	6
tryptophan	acidEt.	27	15.3	− 2.22	6
1,5-naphthalenediol	acidEt.	27	22	− 1.56	6
2,3-naphthalenediol	acidEt.	27	14.9	− 1.90	6
2,7-naphthalenediol	acidEt.	27	6.1	− 1.72	6
2,3-naphthalenediol	acid	27	162		20
1,5-naphthalenediol	acid	27	111		20
2,7-naphthalenediol	acid	27	102		20
adenosine	7	17	120	− 1.6	25
2-aminopurine mononitrate	7	30	6.36 (0.488)		29
hypoxanthine	7	30	3.94 (0.49)		29
6-methylpurine	7	30	0.5 (0.656)		29
purine	7	30	0.18 (0.689)		29
2,4,6-triaminopyrimidine	7	30	5.6		29
4,5-diaminopyrimidine	7	30	4.85		29
2-aminopyrimidine	7	30	2.76		29
thymine	7	30	2.44 (0.51)		29
4-amino-5-aminomethyl- 2-methylpyrimidine	7	30	2.5		29
cytosine	7	30	1.82 (0.60)		29
uracil	7	30	0.525 (0.60)		29
6-azathymine	7	30	0.5 (0.561)		29
6-azauracil	7	30	0.143 (0.71)		29
selenopurine	7		92		31
β-hydroxethylaminopurine	7		85		31
aminopurine	7		82		31
bis(β-hydroethyl)amino purine	7		66		31
hydroxylaminopurine	7		44		31
hydroxypurine	7		37		31
methoxypurine	7		31		31
diethylaminopurineR	7		346		31
dimethylaminopurineR	7		153		31
catechol	acid		3.2		20
hydroquinone	acid		2.9		20
resorcinol	acid		3.3		20
pyrogallol	acid		5.5		20
2-methylresorcinol	acid		5.5		20
phloroglucinol	acid		7.4		20

TABLE 7.1—cont.

Donor	pH	$T(°C)$	$K_c(M^{-1})$	$\Delta H°$ (kcal/mole)	Ref.
4-methylcatechol	acid		6.2		20
orcinol	acid		9.2		20
trimethylhydroquinone	acid		10		20
3,4-dimethylphenol	acid		55		20
2,5-dimethylphenol	acid		88		20
1,4-naphthalenediol	acid		55		20
1,7-naphthalenediol	acid		98		20
3,5-diiodo-4-hydroxy- benzoic acid	8.4	25	42.8		22
idem	alkali	25	10.5		22
3,5-diiodotyrosine	8.4	25	41.3		22
2-amino-6-methylaminopurine	7	30	17.2		29
guanine	7	30	14.2	(0.307)	29
xanthine	7	30	13	(0.397)	
2-amino-6-methylpurine	7	30	8.8		29
8-azaadenine	7	30	8.7	(0.528)	29
8-azaxanthine	7	30	7.9	(0.486)	29
adenine	7	30	7.65	(0.486)	29
2,6-diaminopurine	7	30	6.55	(0.398)	29
6-methylaminopurine	7	30	91.6	~ -9	29
ethylaminopurineR	7	25	101		31
methylaminopurineR	7	25	96		31
n-propylaminopurineR	7	25	95		31
methylmercaptopurineR	7	25	82		31
isopropylaminoR	7	25	80		31
di-n-propylaminopurineR	7	25	79		31
aminopurineR	7	25	77		31
iodopurineR	7	25	57		31
mercaptopurineR	7	25	49		31
hydroxypurineR	7	25	48		31
bromopurineR	7	25	37		31
chloropurineR	7	25	3		31
cytochrome b_5 reductase	8.1	2	5.10^4		47

Coefficient of highest filled molecular orbital shown in parenthesis.
Values from Pullman, B. and Pullman, A. (1963). "Quantum Biochemistry," Interscience, New York.
Coefficient of correlation of K_c vs. coefficient of highest filled molecular orbital $= -0.87 \gg 99.5\%$ confidence level.
R = ribose.

at the 6-position (29, 30, 31). The order of complexing for successive N-alkylation in the 6-amino group of purines is ethyl > methyl > propyl (31) (Table 7.4).

The effect of substitution on the ability of the flavins to form complexes has been examined by several authors. Apart from the work of Harbury and Foley (3) already described, Tsibris, McCormick and Wright (29) have shown that the complexing ability of RFN is decreased by acetylation of the N9 ribityl chain. Chassy and McCormick (30) have looked at the fluorescence quenching as a function of pH of a large number of FMN and FAD analogues. FMN analogues which lack hydroxyl groups in the side chain at the 9-position are more fluorescent than those with glycityl chains which suggests that some self-quenching in the glycityl is due to the secondary hydroxyl groups reacting with the isoalloxazine ring system. Hydroxyl groups in the ribityl portion of the FAD do not appear to be obligatory for the intramolecular complex to be formed. Alkylation of the 3-amino position of FAD seems to have little effect on the intra-molecular complex just as alkylation of FMN has no effect on the intermolecular complex formed with free purines.

TABLE 7.2

Data for riboflavin-5'-phosphate complexes

Donor	pH	$T(°C)$	$K_c(M^{-1})$	$\Delta H°$ (kcal/mole)	Ref.
tryptophan	7		60		4
	7	25	92	− 7.4	7
2,3-naphthalenediol	7	25	24.2		7
5-hydroxytryptophan	7	25	130	− 4.7	7
tryptamine	7	25	106	− 9.2	7
indole-3-acetic acid	7	25	40	− 7.9	7
pyrrole	7	25		− 3.5	7
catechol	acid		32		15
2,3-naphthalenediol	acid		68		20
catechol	6.8		10.4		20
2,3-naphthalenediol	6.8		242		20
tryptophan	6.8	27	98.4		6
2-methylindole	6.8	27	89.6		6
3-methylindole	6.8	27	48.8		6
1,2-methylindole	6.8	27	45.2		6
3-methylindole	acid	27	∼ 60 ∼ 70		6
tryptophan	acid	27	∼ 30 ∼ 40		6
1,4-naphthalenediol	6.8	27	230		6
1,5-naphthalenediol	6.8	27	191		6
2,3-naphthalenediol	6.8	27	176		6
2,7-naphthalenediol	6.8	27	124		
1,7-naphthalenediol	6.8	27	73		6
resorcinol	6.8	27	33.7		6
2,3-naphthalenediol	acid		68		20
catechol	acid		0.68		20
serotonin	6.9		400		4
a-chymotrypsin	5		4.5×10^2		12
partially denatured a-chymotrypsin	5		3.9×10^2		12
cytochrome b_5 reductase	8.1	2	$1.25 . 10^8$		47

Fluorescence quenching studies of both inter and intra molecular complexes involving adenosine show that the fluorescence intensity reaches a maximum at about pH 3 or at least begins to decrease at pH's more acidic

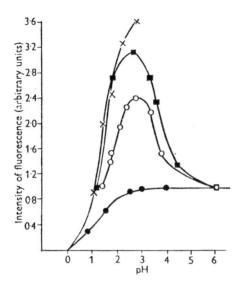

Fig. 7.15. Effect of pH on the fluorescent efficiency of FAD and of riboflavin in the presence of quenchers. $\times - \times$, Riboflavin; $\blacksquare - \blacksquare$, FAD; $\bigcirc - \bigcirc$, riboflavin in 0.05 M adenosine; $\bullet - \bullet$, riboflavin in 0.05 M caffeine. Concentration of riboflavin of the solutions 5×10^{-6} M concentration of FAD is the same. (Determined by the absorption at 454 nm.) Reproduced with permission from Fig. 4, ref. 25.

than this (6, 25, 29, 30, 31). This is explained as due to the protonation of the adenosine amino group at the 6-position whose pKa = 3.3 (30, 31).

Little work has been carried out on the infra-red spectra of flavin complexes. Kyogoku and Yu (32) have examined the infra-red spectra of the 1 : 1 complexes formed between RFN-2′,3′,4′,5′-tetraacetate and tetrabutyrate and 9-ethyladenine in chloroform. The carbonyl bands and the NH bands show slight shifts which though interpreted by the authors as arising from hydrogen bonding can be explained in different terms as discussed later. The association constants of these complexes have been measured from new bands characteristic of the complexes which appear in the infra-red between 3500 and 3300 cm^{-1} (see Fig. 7.16).

Kinetic methods have been used to study the interactions of purines and pyrimidines with flavins. The work of Radda and Calvin on the photobleaching of the flavins and the inhibition of the photoreduction of flavins by NADH of EDTA has already been described in the presence of indoles (13, 14) (Section 7.2). Similar results are noted for purines.

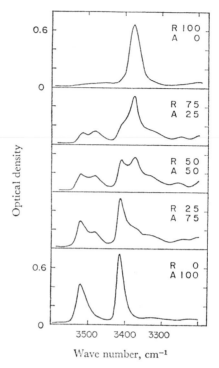

Fig. 7.16. Infra-red spectra of various mixtures of riboflavin tetraacetate (R) and ethyl-adenine (A). The total concentration of the material is 0.02 M and the path length is 2 mm. Reproduced with permission from Fig. 1, ref. 32.

TABLE 7.3

Data for lumiflavin complexes

Donor	pH	$T(°C)$	$K_c(M^{-1})$	$\Delta H°$ (kcal/mole)	Ref.
1,2-dimethylindole	acidEt.	27	284	− 1.26	6
3-methylindole	acidEt.	27	67.1	− 1.99	6
tryptophan	acidEt.	27	28.2	− 2.0	6

The effect of caffeine on the base-catalysed degradation rate of RFN has been studied by Guttman and utilized by him to derive an enthalpy of dissociation of − 5.7 kcal/mole which agrees well with values derived using non-kinetic methods (33).

The interaction of another purine, 8-chlorotheophylline (ClTh) with 9-methylisoalloxazine or 3,9-dimethylisoalloxazine and its effect on their rates of hydrolysis has been studied (34). The effect of the purine on the absorption spectra of the flavins is similar to the changes reported by other workers. The spectral changes are most marked at pH 8 and negligible outside the range pH 4 to 8. As the flavins ionize at the high pH's and ClTh at low pH, the proposed reaction scheme is

$$FlH + ClTh^- \rightleftharpoons (FlH : ClTh^-) \text{ complex.}$$

The association constants from the kinetic and solubility methods do not agree within the experimental error with the constants derived optically from the Benesi-Hildebrand equation. This may however reflect the inadequacies of the Benesi-Hildebrand equation.

The interaction between the excited states of flavins and purines has been demonstrated by Shiga and Piette (16). The triplet lifetime of various flavins are quenched in the presence of adenosine. The triplet lifetime of FAD is

TABLE 7.4

Association constants of complexes of 2-substituted riboflavins and 6-substituted purine ribosides

2-substituted riboflavin	6-substituted riboside purine $K_c(M^{-1})$	
β-hydroxyethylamino	amino	79
	methylamino	110
	dimethylamino	197
methylmercapto	amino	61
	methylamino	134
	dimethylamino	309
morpholino	amino	45
	methylamino	134
	dimethylamino	309
anilino	amino	27
	methylamino	54
	dimethylamino	144

Reproduced with permission from Table 2, ref. 31.

similar to that of the FMN adenosine complex and is much less than the triplet lifetime of FMN. Hence further evidence is given for the existence of an intramolecular complex between the flavin and adenosine moieties.

Although it has been stated that there is little difference between the absorption spectra of complexes of FMN whether with phenols, indoles or purines, certain slight differences have been pointed out which may be significant (7). The difference spectra of FMN indole complexes has peaks and troughs which do lie at slightly different wavelengths to the similar peaks and troughs of FMN purine complexes. The greatest difference is only about

10 nm. Additionally the tryptophan FMN complex has a marked extension of the absorption tail into the red in solution at room temperature. A further difference is that the addition of purines to reduced FMN, i.e. FMNH$_2$,

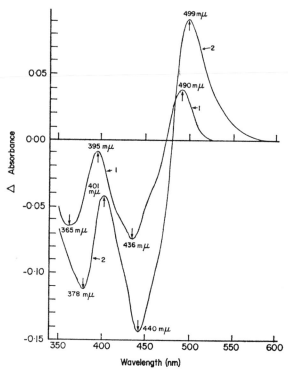

Fig. 7.17. Difference spectra of FMN–adenosine and FMN–tryptophan complexes. Curve 1, plus adenosine; curve 2, plus tryptophan. Reproduced with permission from Fig. 2, ref. 7.

causes an increase in absorption, an effect not observed by the indoles (but see p. 144 in which no change in the effect of the different compounds on the peak shift of RFN is reported (25).

The intramolecular interaction between the flavin and adenine moieties in FAD have been well established by the work of Weber (25). Various studies have been carried out to establish the structure of this molecule. The general concensus appears to be that the two rings are stacked in a parallel mode (29). However a study using circular dichroism techniques (35) has suggested that the two moieties are in close proximity but in a non-planar arrangement. The agreement of experimentally determined spectral energies and rotational strengths with theoretical values computed by Song using the Miles and Urry (35) model is excellent (36).

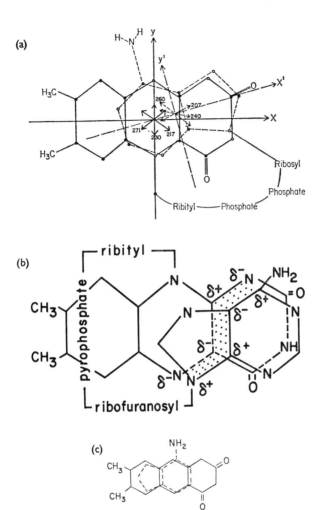

Fig. 7.18. Three suggested structures for FAD. (a) Considered structure for a stacked conformation of FAD. The essential statement of the indicated stacking is the angle made by the planar projection of the x and x^1 axes. Translation of either base may occur along its long axis. The adenine moiety (dashed structure) underlies the isoalloxazine moiety (solid line). Directions of transition moments are similarly indicated. Open circles indicate nitrogens. Reproduced with permission from Fig. 3, ref. 35. (b) Reproduced with permission from ref. 29. (c). After ref. 36.

7.5 Aromatic Hydrocarbons

The interaction of aromatic hydrocarbons with flavins was first observed by Sakai (37) who reported on the solubilization of flavins by hydrocarbons and on the quenching of flavin fluorescence by the same hydrocarbons.

More recently, the complexes formed between polycyclic aromatic hydrocarbons and RFN have been studied by McCormick *et al.* (38). The addition of the hydrocarbons causes slight changes in both the absorption and fluorescence spectra of RFN. From the absorption changes as a function of hydrocarbon concentration it is inferred that 1 : 1 complexes are formed. The association constants evaluated by the Benesi-Hildebrand equation show a correlation with the ionization potentials of the hydrocarbons, -0.68 at 96% confidence level (Table 7.5).

TABLE 7.5

Association constants of complexes of riboflavin and aromatic hydrocarbons in absolute ethanol

Hydrocarbons	$K_c(M^{-1})^a$	Ionization potential (eV)[b]
azulene	438	
acenaphthalene	112	
1,2-benzanthracene	82	7.45
benzo(g,h,i)fluoranthene	60	
1,2-benzofluorene	56	
pyrene	53	7.55
anthracene	53	7.37
triphenylene	53	8.09
2,3-benzofluorene	47	
fluoranthene	36	7.72
phenanthrene	28	8.09
naphthalene	22	8.10

Correlation coefficient $= -0.68$ at 96% confidence level.
[a] Ref. 38.
[b] Slifkin, M. A. (1963). *Nature*, **200**, 877.

7.6 Coenzymes

Mixtures of NADH or NADPH with FMN show marked increase in colouration on freezing down to $-78°C$ (39). The absorption spectra of these mixtures show the same differences to the spectrum of free FMN as does tryptophan FMN mixtures. A solid complex formed from NADH and FMN gives an ESR signal presumed to be the FMN free radical. Although this signal was obtained in the dark, the sample was made up in the light (40).

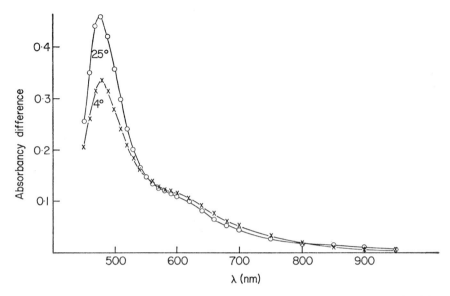

Fig. 7.19a. Absorption spectrum of mixture of FMNH$_2$ and NAD$^+$ at 4° and 25° in pH 6.3 buffer (10 M of NAD$^+$ to 1 M of flavin). Reproduced with permission from Fig. 9, ref. 48.

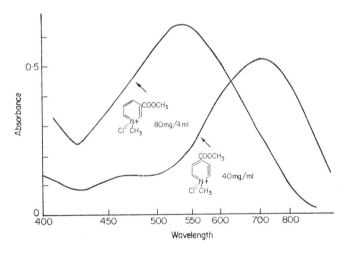

Fig. 7.19b. Charge-transfer bands of the complexes between FMNH$_2$ and some NAD$^+$ analogues. Concentrations are: FMNH$_2$, 1.5 mM and (a) N-methyl-3-carbomethoxy-pyridinium chloride 20 mg/ml and (b) N-methyl-4-carbomethoxypyridinium chloride 40 mg/ml. Solvents are phosphate buffer. Reproduced with permission from Fig. 1, ref. 39.

The work of Radda and Calvin (13, 14) on the inhibition of the photoreduction of flavins by NADH has been referred to earlier.

Massey and Palmer (48) observed new long wavelength bands in mixtures of reduced FMN, i.e. $FMNH_2$ and either NAD^+ or N-methylnicotinamide. These new bands showed temperature reversibility.

Sakurai and Hosoya (41) have confirmed the work of Massey and Palmer showing that NAD^+ and its analogues form 1 : 1 complexes with FMN_2. The mixtures are strongly coloured due to the presence of the new broad structureless band whose position is a linear function of the lowest unoccupied molecular orbital energy, i.e. the electron affinity, of the NAD analogues. The association constants of the complexes which were obtained with the Benesi-Hildebrand equation do not correlate with the lowest unoccupied molecular orbital energies (see Fig. 7.10).

Proton magnetic resonance has been used as a tool to investigate the interaction between NAD analogues and FMN as a means of investigating

TABLE 7.6

Data for isoalloxazine complexes

Method of Determination	$K_c(M^{-1})$ 9-methylisoalloxazine 8-chlorotheophylline	3,9-dimethylisoalloxazine 8-chlorotheophylline
kinetic	92.5	98
solubility[a]	88.5	....
spectral	52	48.6

Adapted from Table 4, ref. 34. [a] Determined at 37°C; all other measurements taken at 35°C.

the intramolecular interaction between the flavin and adenine moieties in FAD (42). The data is interpreted as showing that in the dinucleotide the benzene protons of the isoalloxazine ring are strongly shielded by the adenine ring thus indicating a strong intramolecular interaction between the two groups.

7.7 Dimerization

Beinert (43) in his study of the oxidation of flavins, observed concentration dependent absorptions in the near infra-red at 900 nm which he ascribed to the formation of a semiquinone dimer occurring in both alkaline and neutral solutions.

Massey and Palmer (47) after a study of the temperature and concentration dependence of the flavin spectra believe this band to arise from a charge

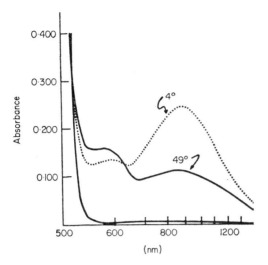

Fig. 7.20. Absorption spectra of 1.1×10^{-3} M FMN at the semiquinoid oxidation level in 0.17 M citrate, pH 6.1; light path 1.00 cm, temperature 4° and 49° as indicated, reduced with $Na_2S_2O_4$. Lower solid line, oxidized form at 4° and 49°. Reproduced with permission from Fig. 4, ref. 43.

transfer complex between FMN and $FMNH_2$ a view reiterated by Gibson *et al.* (45). This complex is similar to the well-known quinhydrone a complex formed between benzoquinone and hydroquinone often cited as a classic example of a charge transfer complex.

Holmström (44) has evaluated the kinetics of the formation of this complex or dimer during photoreduction studies of FMN. The reaction assumed to be of the form $2FMNH \cdot \rightleftharpoons FMN + FMNH_2$ (i.e. after Beinert) has forward rate constants of $3.10^8 M^{-1} sec^{-1}$ in neutral solution and $1.10^8 M^{-1} sec^{-1}$ in alkaline solution. The respective back rate constants are 10^6 and $< 10^5 sec^{-1}$.

Additional evidence for the dimer origin of this band comes from the kinetic studies of Swinehart (46) using the temperature-jump technique. The primary reaction in pH 4.7 buffer is shown to be that suggested by Beinert and Holmström. The enthalpy of dissociation of the dimer is -9.3 kcal/mole. Swinehart states that this reaction should be considered as a hydrogen transfer process rather than an electron transfer process.

Isenberg and Szent-Györgyi (39) have attributed the shoulder at 485 nm which they see in the spectrum of highly concentrated FMN as arising from an FMN self complex. This complex consists of two oxidized FMN molecules held together by charge transfer. Neither the concentration nor temperature dependence of this supposed complex has been established. However circular dichroism studies suggest that association does take place both in FMN and FAD by vertical stacking of the isoalloxazine rings (42).

7.8 Flavoproteins

Strittmatter (47) has investigated the nature of flavin binding in the flavoprotein cytochrome b_5 reductase. FMN binding to this enzyme is severely inhibited in the presence of strong electron acceptors such as trinitrobenzsulphonate and iodine. The reductase also forms complexes with NADH and less strongly with NAD^+. Association constants of the complexes with NADH, NADPH and their derivatives have been determined from fluorescence quenching studies and from competitive binding studies. These association constants are orders of magnitude higher than those for free flavins.

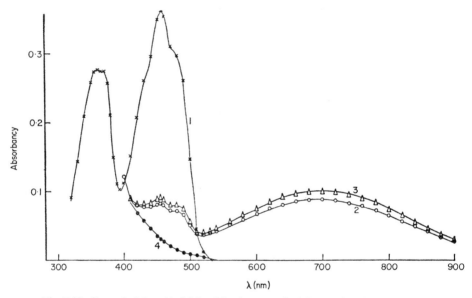

Fig. 7.21. Curve 1, 3.2×10^{-5} M oxidized enzyme in 0.06 M phosphate, pH 6.3, plus 10^{-3} M sodium arsenite, temperature, 25°; curve 2, plus 3.3×10^{-4} M NADH; curve 3, after the further addition of 1.67×10^{-4} M NAD^+; curve 4, after the further addition of NADase (1400 Kaplan units). Reproduced with permission from Fig. 1, ref. 48.

The interaction of the flavoprotein lipoyl dehydrogenase with the coenzymes NADH, TNADH, NHDH and NADPH has been described by Massey and Palmer (48). The spectrum of the enzyme exhibits long wavelength bands which are associated with conditions conducive to full reduction of the enzyme. This band is dependent on the presence of NAD^+ as the effect of adding oxidized coenzymes on the spectrum of the reduced coenzyme is to cause the appearance of the new long wavelength band which gives this enzyme its characteristic green colour. These new long wavelength bands are designated as charge transfer bands with reduced enzyme as donor (Table 7.11).

Burton (26) has shown that the binding of FAD to the enzyme D-amino acid oxidase is competitive with various purines. Some of the competing purines such as caffeine inhibit the enzyme but purines related to FAD such as AMP whilst strongly competing with FAD in fact protect the enzyme. The inhibition of the enzyme by some antagonists in the presence of FAD is thought to be due to the formation of non-fluorescent complexes between the inhibitors and FAD.

Yagi and Ozawa (49) have examined the competitive binding of flavins and purines to the enzyme apo-D-amino acid oxidase and found that both flavin and purine compete with FAD but not with each other, suggesting that both the flavin and adenosine moieties of FAD bind to the enzyme.

Yagi et al. (18) have found that p-aminosalicylic acid inhibits the activity of the enzyme by forming a complex with FAD, by competitive inhibition with FAD for the enzyme and by competitive inhibition with the substrate. Similar results have been found for a large number of phenols (Section 7.3).

Studies of the binding of flavin phosphates to some FMN-dependent enzymes have been carried out by Tsibris, McCormick and Wright (50). From inhibition or reactivation studies of the enzymes or flavins, they show that there is no specific dependence on the presence of groups at special sites in the alloxazine rings but suggest that the whole or at least a portion of the ring takes part in binding. The aromatic amino acid residues of the enzyme protein may also be concerned in binding to the flavin.

Veeger et al. (51) have looked at the spectral changes induced on the addition of various compounds, all of which are electron-donating, to the flavoproteines, succinate dehydrogenase, D-amino acid oxidase and L-amino acid oxidase, and have stated that the changes arise from the formation of charge transfer complexes.

The interaction of D-amino acid oxidase with various aliphatic amino acids is described in Section 3.11. Currently with these studies Yagi has investigated the interaction of the enzyme and its purple complex with amino acids, with sodium benzoate (64). The spectrum of the enzyme benzoate complex is the same as the spectrum arising from the interaction of the purple complex with benzoate (Fig 7.22). Old purple complex which is believed to change to the semiquinoid form of the enzyme (see Section 3.11) shows only a partial change to the spectrum of the enzyme benzoate complex and there is a blue colouration. This is ascribed to the formation in some part to a semiquinoid enzyme benzoate complex which however reverts fully to the enzyme benzoate complex on the introduction of oxygen. There exists isosbestic points in the spectra during conversion of the purple complex with benzoate which is interpreted as being due to the benzoate complexing with the enzyme by displacement of the amino acid. Other compounds giving spectra on addition to the enzyme similar to those given by sodium benzoate are D-lactate and

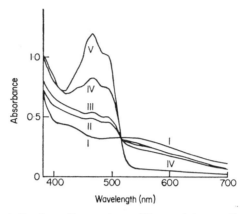

Fig. 7.22. Change in the absorption spectrum of the purple intermediate upon mixing with known concentrations of sodium benzoate. (I) The purple intermediate ($1.06.10^{-4}$ M in respect to FAD). The amounts of sodium benzoate added to this solution were, in final concentrations, (II) $5.0.10^{-3}$ M; (III) $1.0.10^{-2}$ M; (IV) $5.0.10^{-2}$ M; and (V) $1.0.10^{-1}$ M. All operations were made under anaerobic conditions. Reproduced with permission from Fig. 8, ref. 64.

D-mandelate (65). It should be emphasized that these spectra are very similar to the vast majority of spectra observed on adding electron donors to flavins. Changes in the circular dichroism of the enzyme on the addition of these compounds is attributed also to complex formation.

The interaction of various carboxylic acids with D-amino acid oxidase have also been undertaken by Yagi and his coworkers (66). The addition of these acids causes perturbations of the flavin spectrum which is very similar to changes caused by amino acids containing hydrophobic side chains, for example L-leucine, and L-valine. Association constants of the various carboxylic acid enzyme complexes have been evaluated and shown to depend on the carbon number of the acid up to about 4. The association constants are found to increase if the carboxylic acids contain a C = C bond at the α-β positions. The addition of aminobenzoates to the enzyme causes changes similar to that of sodium benzoate with the exception of o-amino benzoate which causes the appearance of a diffuse band at about 550 nm. Some other carboxylates namely Δ^1-piperidine-2-carboxylate and pyrrole-2-carboxylate give similar results. It is thought that these compounds behave as n-electron donors to the enzyme. The reason why these latter chemicals can act as n-donors and not the L-amino acids or p-aminobenzoate is believed to be because of the orientation of the orbitals of the lone pair electrons on the two nitrogens vis-a-vis the position of the carboxylate group. The optical rotary spectra of the mixtures also show the similarity between the interaction of o-aminobenzoate acid and the carboxylates (see Tables 7.12 and 7.13).

7.9 Interaction Mechanisms

There are many problems to be answered concerning the interaction of flavins. Certainly very many kinds of nucleophilic compounds can form complexes with the flavins. The nature of these complexes is however the subject of much controversy. The first is what is the cause of the spectral changes of the flavins in the complexes.

7.9.1 STABILIZATION OF THE SEMIQUINONE

In almost all cases involving the oxidized flavin, the same changes occur. The differences pointed out by Wilson (7) between different types of compounds are very slight and in the writer's view are not significant but merely represent a difference in degree of interaction rather than kind. Isenberg and Szent-Györgyi (4) ascribe the new spectrum to that of the semiquinone mainly because the position of the new absorption in the complex at about 500 nm is the position where the semiquinone, at least in acid solution, is known to absorb (43). The writer (27, 52) has also shown that in addition to the new absorption at 500 nm there is an isosbestic point at 330 nm which indicates that the spectrum isn't just some form of perturbed spectrum of the free flavin, as proposed by Kosower (53) but represents a stage in the process in reduction of the flavin, as shown by Beinert (43). Although Kosower has stated that the broadening of the flavin spectrum gives rise to an apparent band at 500 nm in the difference spectrum (this is illustrated in Fig. 2.2), the spectra of some of the stronger complexes show quite clearly resolved bands at 500 nm even when using absorption spectra (27, 28). It has also been demonstrated that the addition of the complexing reagents protect the flavins against complete reduction (27, 52) which again indicates that the semiquinone is stabilized in the complex, it being remembered that the reduction of flavin is a two-step process

$$FlH \rightleftharpoons FlH_2 \cdot \rightleftharpoons FlH_3 \qquad \text{ref. 43}$$

Although it is postulated that the spectrum of complexes oxidized flavin is that of the semiquinone, it does not follow that the major binding forces arise primarily from charge transfer. It can be envisaged that two components in a complex are held together by unspecified forces thus allowing the electron to pass over to the flavin to produce the spectrum of the semiquinone.

7.9.2 STRUCTURE OF COMPLEXES

Good evidence for the role of charge transfer forces comes from the structure of the complexes between flavin bromides and hydroquinone (61) and naphthalene diol (63). Not only does overlap occur between regions of high and low electron density, but in the case of the hydroquinone complex where the intramolecular distance has been measured, there is agreement with the distances found by Wallwork for the intramolecular distances in hydrocarbon charge transfer complexes. The flavins in these complexes being bromides are probably in the semiquinone form before complexing (23) and hence presumably poorer donors than fully oxidized flavins. Consequently it can be assumed that charge transfer is even more important in complexes of fully oxidized flavins.

It should be emphasized that both hydroquinone and naphthalene diol are good electron donors as they both possess two electron donating hydroxyl groups. Hence other weaker electron donors may not necessarily form charge transfer complexes with flavins. An interesting point is that the structures of the two complexes are different involving as they do different regions of the isoalloxazine ring. This suggests the generality of charge transfer interactions in flavin complexes as various geometrical approaches to the flavins are available to the complexer and steric hindrance is not too important.

7.9.3 ESR OF COMPLEXES

No identifiable ESR signals have been observed when complexes have been prepared and examined in the dark. However, as discussed in Section 2.1.6.1, the processes giving rise to ESR signals in complexes are not fully understood and it is not certain that the detection of signals would be confirmation for the charge transfer nature of the complexes.

7.9.4 CORRELATION STUDIES

Some indication of charge transfer stabilization in the ground state of the complexes is given by the different correlation studies. For example, there is a good correlation between the energies of the highest occupied molecular

TABLE 7.7

Association constants of benzene derivatives with FAD or the oxidase protein

	Direct complexing with FAD $K_c(M^{-1})$	Competitive inhibition with FAD or the protein		Competitive inhibition with D-alanine	
		Method I	Method II	Method I	Method II
phenol	31.2	17.5	17.8		
salicylic acid	1540	1000	910	2860	2860
m-aminophenol	9.1	38.4	33.3		
benzoic acid				71500	77000
p-aminobenzoic acid		139	137	3220	3330
aniline		109	110		
p-nitrobenzoic acid		1090	1280	192000	244000

Reproduced with permission from Table 1, ref. 18.

orbitals of the purines and pyrimidines and the association constants of their complexes with flavins (29). In addition we might note that those purines containing the substituted amino group at the 6-position form particularly strong complexes, the strength of which increases with increasing methylation

TABLE 7.8

Association constants of FAD complexes

Compound	$K_c(\mathrm{M}^{-1})$
adenosine triphosphate	25.6
adenosine diphosphate	27
adenosine-5′-phosphate	25
adenosine	33.4
caffeine	100
quinine	370

Adapted from Table 4, ref. 26. In pH 8.3 buffer at 17°C.

of the amino group (31). It has been shown that the phenols form fairly strong complexes with flavins whereas the equivalent benzene derivatives or alternatively phenols containing electron-withdrawing groups do not complex under similar conditions (19, 20).

There is significant correlation between the ionization potentials of the hydrocarbons and the association constants (38).

7.9.5 PURINES AS n- AND π-DONORS

There appears to be two different mechanisms operating in the complexing of purines with flavins. Complexing occurs either by π-donation from the purine ring or by n-donation from the nitrogen at the 6-position. Some evidence for n-donation comes from the infra-red studies of the RFN-9 ethyladenine complex which shows carbonyl band shifts (32) associated either with hydrogen bonding or with charge transfer complexing. Hydrogen bonding can be ruled out as a general mechanism operating between these substituted purines and flavins as increased alkylation of the amino group increases the complexing ability whilst at the same time removing the possi-bility of hydrogen bonding (31). Although the purines also possess n-electrons on the heterocyclic nitrogen atoms, it is not considered that these take part in complexing as the lone-pair orbitals are in the plane of the ring and it would be necessary for the purine and flavin to be perpendicular for which there is no evidence.

7.9.6 CHARGE COMPLIMENTARITY

Another form of interaction which can be thought of as a localized charge transfer interaction occurs when electronegative parts of one molecule are aligned with electropositive parts of another. Such an interaction has been suggested for purine complexes with indole (54) and flavins (29). An example is shown in Fig. 7.18b. This would explain why correlations between ionization potentials and association constants were only approximate as theoretical ionization potentials are usually those of the π-electron systems and not those of the n-electrons or other localized regions of charge. The ionization potentials of the conjugated system, will however, have an influence on the ionization potentials of localized regions.

7.9.7 DIFFERENCES BETWEEN PURINES AND INDOLES

It was pointed out that there are apparent differences between the interactions of purines and indoles with flavins (7). The addition of purines to reduced flavins causes an increase in absorbance an effect not shown by the indoles. Wilson (7) considered that the reduced flavins might complex as donors with purines but dismissed this on the grounds that the purines are poor electron acceptors. Terms such as donor and acceptor are of course very relative and there does not seem any *a priori* reason why the purines should not be electron acceptors with reduced flavins, which can act as donors with NAD^+ (41). There is also evidence that pyrimidines are electron acceptors in complexes with the moderate donors, benzene and toluene (56). By extension as the purines are thought to be better electron acceptors than the pyrimidines (62), then the concept of purines behaving as electron acceptors towards reduced flavins is quite acceptable.

7.9.8 INDOLE COMPLEXES

As well as the changes which have been ascribed to the semiquinone, the indole flavin complexes exhibit at room temperature, new bands specific to the indole used. However, as no correlation with ionization potential is found, it is difficult to accept that these are charge transfer bands, unless the charge transfer is very localized being completely uninfluenced by the conjugated ring system. Karreman, Isenberg and Szent-Györgyi (54) have suggested that these long wavelength bands might arise from a reverse charge transfer interaction in which the flavin donates to the indole. Certainly if the flavin in the complex is in the semiquinone form then this should be a better donor than when in the oxidized form. On the other hand no interaction was observed by Wilson (7) between fully reduced flavin, an even better donor, and the indoles.

7.9.9 Phenol Complexes

The bands observed in pheno-flavin complexes correlate with the ionization potentials of the flavins. This is indicative of their being charge transfer bands. Kosower (53) has expressed doubts about the validity of the calculations used to obtain the ionization potentials.

TABLE 7.9

Association constants of phenol and nitrophenols with FAD or the oxidase protein

	Direct complexing with FAD	Competitive inhibition with FAD or the protein
phenol	31.2	17.5
p-nitro	2×10^3	2×10^2
2,4-dinitro	5×10^4	2.56×10^2
2,6-dinitro	2.5×10^4	4.35×10^2
2,4,6-trinitro	8.35×10^4	1.56×10^3

Reproduced from Table 1, ref. 18.

7.9.9.1 *Tyrosines*

The tyrosines are an anomolous group of phenols. Increasing halogenation increases their complexing ability with flavins (21, 22). The halogens are strong electron-withdrawing groups and if, as appears to be so, the oxidized flavins are electron acceptors, a view borne out by the increased association constants in acid solution as compared with alkali solution, then this is an unexpected result. A possible explanation is that the tyrosines are n-electron donors to flavins just as they are to chloranil (55) and so therefore the substitution of electron withdrawing groups into the phenol ring does not have the same effect as it must have if the tyrosines are π-donors.

7.9.10 Validity of Association Constants

In the above and prior discussions, the association constants have been assumed to give directly the relative strengths of the complexes. Also inter-comparison has been made between association constants obtained from absorbance studies and fluorescence studies. It is doubtful whether association constants are a good measure of complex stability as discussed in Sections 1.6 and 1.7, except under special conditions, i.e. for closely related groups of molecules in the same solvent and pH if in water, and at the same temperature. As previously mentioned, complexing can take place with excited molecules so that whereas absorbance measurements given an indication of the strength of ground state complexing, fluorescence derived values may contain a contribution from an excited state complex (13).

7.9.11 Hydrogen Bonding

Although a case has been made out for the existence of charge transfer forces in certain flavin complexes, other forces cannot be excluded. One very common force in biomolecular systems is hydrogen bonding. The earliest workers attached some importance to the existence of the 3-amino group of isoalloxazine thinking that hydrogen bonding at this site was responsible for complex formation. It has subsequently been established that substitution of other groups into this position removing the possibility of hydrogen bonding does not affect complex formation (3, 50, 57). Hydrogen bondings at other sites can coexist with charge transfer. In the riboflavin dihydrobromide hydroquinone complex bromide ions are hydrogen bonded both to the flavin and the hydroquinone (61).

7.9.12 Semiquinone Flavin Complexes

The interaction of flavin with HI produces a semiquinone flavin which can then interact with electron donors to form complexes (23). Presumably the flavin dihydrobromide complexes with electron donors (61, 63) also involve the semiquinone. As in the case of the oxidized state the semiquinone acts as an electron acceptor.

7.9.13 Reduced Flavin Complexes

The fully reduced flavins also form complexes. The complexes with NAD^+ is a well-authenticated case of charge transfer complexing in the sense as defined by Mulliken. Charge transfer bands show a clear correlation between the position of the bands and the electron affinity of the different coenzymes (41). It can be presumed that the changes in spectrum seen in adding purines to reduced flavins is caused by a similar interaction (7). In this context, Matsunaga (58) has shown that the addition of both conventional organic charge donors and acceptors to the flavin analogue 1,3-dimethylisoalloxazine produces colour changes. Solids have been isolated from the solutions containing isoalloxazine and donor or acceptor in 1 : 1 stoichiometry. This suggests that the 1,3-dimethylalloxazine forms charge transfer complexes both as a donor and an acceptor. It is not surprising if the flavins operate in a similar mode.

7.9.14 Solvent Effects

The role of the solvent must also be considered in discussion of charge transfer complexing.

Harbury et al., have examined the absorption spectra of 3-methyl lumiflavin in a large number of solvents, many of which are potential charge donors of hydrogen bonding solvents (59). Differences in spectra are observed which are

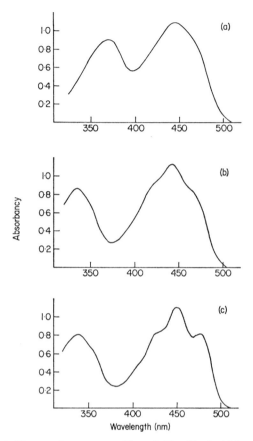

Fig. 7.23. Solvent effects on the spectrum of 3-methyl lumiflavin. (a) Spectrum of 3-methyl-lumiflavin in water, pH 7.3; 1.9×10^{-5} M solution, 5 cm cuvette. (b) Spectrum of 3-methyl-lumiflavin in dioxane; 1.9×10^{-5} M solution, 5 cm cuvette. (c) Spectrum of 3-methyl-lumiflavin in benzene; 1.9×10^{-5} M solution, 5 cm cuvette. Reproduced with permission from Fig. 1, ref. 59.

thought to reflect the degree to which hydrogen bonding to the flavin plays a role in solvent flavin interaction. The authors found that interaction with charge donors but not hydrogen donors was greatly increased when the flavin was in a hydrogen bonding solvent, a result subsequently confirmed by later workers (6, 7, 20). It must be interpolated that the hydrogen bonding solvents are also the more polar solvents which could provide an answer as to why charge transfer interaction is facilitated in these solvents. There are changes in the spectrum of a flavin in the highly polar but non-hydrogen bonding solvent dimethyl sulphoxide on the introduction of charge donors (5) although these are somewhat different to those normally observed.

TABLE 7.10

Properties of the complexes $FMNH_2-NAD^+$ analogues and the lowest vacant (LV) and highest occupied (HO) molecular orbital energies of the NAD^+ analogues

	R in N-methyl-pyridinium chloride	λ_{max} (mμ)	$\varepsilon_{M,C}^{max}$	K (M^{-1})	LV^a ($-\beta$)	HO^a ($-\beta$)
I	3-$CONH_2$	527	830	4.3	0.34 (0.34)	0.56 (0.56)
II	4-$CONH_2$	655	600	5.0	0.16 (0.17)	0.58 (0.58)
III	3-$COOCH_3$	540	1000	7.7	0.34 (0.34)	0.64 (0.64)
IV	4-$COOCH_3$	710	1050	8.8	0.15 (0.16)	0.66 (0.66)
V	3-COOH	455	1200	2.5	0.34 (0.35)	0.77 (0.77)
VI	4-COOH	455	2500	0.9	0.16 (0.17)	0.79 (0.78)
VII	3-$COCH_3$	575	80	7.6	0.30 (0.31)	1.21 (1.21)
VIII	4-$COCH_3$	800	—^{b}	—^{b}	0.06 (0.08)	1.16 (1.16)
IX	3-NH_2	530	300	4.1	0.36 (0.37)	0.36 (0.36)
X	4-NH_2	—^{c}	—^{c}	—^{c}	0.48 (0.48)	0.44 (0.44)
XI	H	540	600	1.1	0.36 (0.36)	1.16 (1.16)
XII	ADPR^d	455	560	11.4	0.34 (0.34)	0.56 (0.56)

^{a} Calculated without (with) inclusion of hyperconjugation effect of the methyl groups.
^{b} Gradually decomposes.
^{c} Cannot form complex with $FMNH_2$.
^{d} Represents the adenine–phosphate–ribosyl moiety of NAD^+.

Coefficient correlation of λ_{max} vs. $LV = -0.93$ at 99.5% confidence level.
No other parameters show any significant correlation.
Reproduced with permission from Table 11, ref. 41.

A recent study of the spectra of RFN and lumiflavin in different solvents leads to the conclusion that the flavins are solvated in the more polar solvents, the bonding being between the solvent and the isoalloxazine ring of the flavin (60). This bonding could be either hydrogen bonding or the formation of a charge transfer complex. The spectrum of RFN in the polar charge donating but non-hydrogen bonding solvent dimethyl sulphoxide (5) is of the same spectral type as dioxane, a non-polar non-hydrogen bonding solvent (60).

There is a different interaction between the hydrogen bonding solvents and the non-hydrogen bonding but polar solvents with flavins which suggests that solvation does occur only through hydrogen bonding. Harbury et al. (59) point out that a hydrogen bonded flavin will be a better electron acceptor than the unbonded flavin as hydrogen bonding leaves the isoalloxazine ring

TABLE 7.11

Association constants of cytochrome b_5 reductase nucleotide complexes

Nucleotide	$K_c(M^{-1})^a$
NADH	$2.5 . 10^6$
3-acetylpyridine-NADH	$6.7 . 10^6$
NAD	$9 . 10^4$
ADP-ribose	$\sim 5 . 10^5$

In pH 8 buffer at 5°.
[a] Ref. 47.

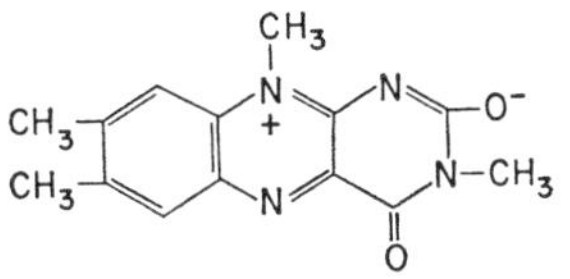

Fig. 7.24. Structure of hydrogen bonded lumiflavin showing increased electropositivity in the isoalloxazine ring. After ref. 59.

more electropositive. Whilst hydrogen bonding aids donation to the flavin there is the possibility that a polar environment also helps as the interaction of charge donors with lumiflavin in dimethyl sulphoxide does not occur in the non-polar, non-hydrogen bonding solvent benzene (5).

7.9.15 FAD

FAD the prosthetic group of the flavoproteins exhibits both inter and intramolecular complexing. The evidence for this discussed earlier indicates that the intramolecular complexing between the adenosine and flavin moieties are identical to the intermolecular complexing between a flavin and adenosine. Therefore the explanation put forward to explain those complexes apply equally to the FAD intramolecular complex. It would be true to say that the concensus of modern opinion holds that intramolecular bonding in FAD arises mainly from hydrophobic bonding (35, 36, 42).

In intermolecular complexes of FAD there is evidence for both donor and acceptor properties of the flavin. On one hand FAD competes with purines in binding to flavoproteins (49) and complexes very strongly with good charge acceptors such as 2,4,6-trinitrophenol (18), on the other hand it also complexes with a good electron donor chlorpromazine (Section 9.5.1). There is evidence to suggest that the association of FAD into self complexes is via bonding of adjacent isoalloxazine rings (35). There is however no clear indication of how it associates with other molecules, whether via the adenosine or flavin moieties.

7.9.16 FLAVOPROTEINS

The flavoproteins must be considered separately from the free flavins and FAD. Some of the interaction described herein have obvious analogies with the interactions of the free flavins although in general association constants of the protein complexes are very much higher than those of the free flavins,

TABLE 7.12

Association constants of some inhibitors of D-amino acid oxidase

Carboxylate	$K_c(M^{-1})$
n-butyrate	$4.75 \cdot 10^2$
crotonate	$1.43 \cdot 10^4$
hydrocinnamate	$3.22 \cdot 10^2$
cinnamate	$2 \cdot 10^4$

In pH 8.3 buffer at 25°C.
Reproduced from Table 1, ref. 66.

suggestive of additional bonding mechanisms. The protein interaction is in general difficult to explain. The purple complex seen in the interaction of certain amino acids with D-amino acid oxidase is the only complex of a fully oxidized flavin which is a proven charge transfer complex as defined by Mulliken. The aliphatic amino acids and amines are known to be good electron donors (55). Yagi and his coworkers have established a good correlation between the positions of the new broad bands of these complexes with the ionization potentials of the amino acids. The complexing comes about via the lone-pair electrons on the nitrogen atoms in the amino part of the amino acids. This is the only well-attested case of a charge transfer band occurring in complexes of fully oxidized flavins albeit in the flavoprotein form. The appearance of these charge transfer bands is accompanied by the almost total loss of typical flavin absorption. These complexes are obviously

TABLE 7.13

Peak positions of charge transfer bands of complexes of D-amino acid oxidase and carboxylates

Carboxylate	$h\nu_{CT}(eV)$
o-aminobenzoate	2.19
N-methyl o-aminobenzoate	2.14
Δ^1-pyrroline-2-carboxylate	2.02
Δ^1-piperidine 2-carboxylate	1.97

Reproduced from Table 2, ref. 66.

not the same kind of complex observed with flavins not bound to protein. These complexes are produced in the middle of a complicated chain of chemical reactions during which, before interaction to form the purple complex the amino acids are chemically changed. Furthermore in the complex the amino acid is additionally bonded to the protein. Although these complexes are formed during the reduction of the flavoprotein from the fully oxidized to the fully reduced state, they do not involve the semiquinone although on standing they do dissociate to give the semiquinone ion.

Other complexes of the benzoate type formed with this enzyme are similar to those formed by flavins and are probably explicable in the same terms.

In view of the very complicated structures of the flavoproteins and their high chemical specificity, it is not possible to explain the interaction of the flavoproteins without knowing the detailed structure of the flavoprotein and the nature of the prosthetic groups near to the interacting sites. It is unlikely that one general explanation would cover the behaviour of the flavoprotein complexes.

REFERENCES

1. Harte, R. and Chen, J. (1949). *J. Pharm. Sci.* **38**, 568.
2. "Martindale Extra Pharmacopoeia," (1967). The Pharmaceutical Press.
3. Harbury, H. A. and Foley, K. A. (1958). *Proc. natn. Acad. Sci.* **44**, 662.
4. Isenberg, I. and Szent-Györgyi, A. (1958). *Proc. natn. Acad. Sci.* **44**, 857.
5. Slifkin, M. A. Unpublished work.
6. Pereira, J. F. and Tollin, G. (1967). *Biochim. biophys. Acta,* **143**, 79.
7. Wilson, J. E. (1966). *Biochemistry,* **5**, 1351.
8. Wright, L. D. and McCormick, D. B. (1964). *Experientia,* **20**, 501.
9. Isenberg, I., Szent-Györgyi, A. and Baird, S. L. (1960). *Proc. natn. Acad. Soc.,* **46**, 1307.
10. Fory, W., MacKenzie, R. E. and McCormick, D. B. (1968). *J. theoret. Chem.* **5**, 625.
11. MacKenzie, R. E., Fory, W. and McCormick, D. B. (1969). *Biochemistry,* **8**, 1839.
12. Swinehart, J. H. and Hess, G. P. (1965). *Biochim. biophys. Acta,* **104**, 205.
13. Radda, G. K. (1966). *Biochim. biophys. Acta,* **112**, 448.
14. Radda, G. K. and Calvin, M. (1963). *Nature,* **200**, 464.
15. Fox, J. L. and Tollin, G. (1966). *Biochemistry,* **5**, 3865 and 3873.
16. Shiga, T. and Piette, L. H. (1964). *Photochem. Photobiol.* **3**, 213; (1965) **4**, 769.
17. Yagi, K. and Matsuoka, Y. (1956). *Biochem. Z.* **328**, 138.
18. Yagi, K., Ozawa, T. and Okada, K. (1959). *Biochim. biophys. Acta,* **35**, 102.
19. Fleischman, D. E. and Tollin, G. (1965). *Biochim. biophys. Acta,* **94**, 248.
20. Fleischman, D. E. and Tollin, G. (1965). *Proc. natn. Acad. Sci.,* **53**, 38.
21. Cilento, G. and Berenholc, M. (1963). *J. phys. Chem.,* **67**, 1159.
22. Cilento, G. and Berenholc, M. (1965). *Biochim. Biophys. Acta,* **94**, 271.
23. Fleischman, D. E. and Tollin, G. (1965). *Proc. natn. Acad. Sci.,* **53**, 237.
24. Ray, A., Guzzi, A. V. and Tollin, G. (1965). *Biochim. biophys. Acta,* **94**, 258.
25. Weber, G. (1950). *Biochem. J.,* **47**, 114.
26. Burton, K. (1951). *Biochem. J.,* **48**, 458.

27. Slifkin, M. A. (1965). *Biochim. biophys. Acta*, **103**, 365.
28. Slifkin, M. A. (1965). *Biochim. biophys. Acta*, **109**, 617.
29. Tsibris, J. C. M., McCormick, D. B. and Wright, L. D. (1965). *Biochemistry*, **4**, 504.
30. Chassy, B. M. and McCormick, D. B. (1965). *Biochemistry*, **4**, 2612.
31. Roth, J. A. and McCormick, D. B. (1967). *Photochem. Photobiol.*, **6**, 657.
32. Kyogoku, Y. and Yu, B. S. (1969). *Bull. chem. Soc. Jap.*, **42**, 1387.
33. Guttman, D. E. (1962). *J. Pharm. Sci.*, **51**, 1162.
34. Wadke, D. A. and Guttman, D. E. (1965). *J. Pharm. Sci.*, **54**, 1293.
35. Miles, D. W. and Urry, D. W. (1968). *Biochemistry*, **7**, 2791.
36. Song, P-S. (1969). *J. Am. chem. Soc.*, **91**, 1850.
37. Sakai, K. (1956). *Nagoya, J. med. Sci.*, **18**, 237.
38. McCormick, D. B., Li, H-C. and MacKenzie, R. E. (1967). *Spectrochim. Acta*, **23A**, 2353.
39. Isenberg, I. and Szent-Györgyi, A. (1959). *Proc. natn. Acad. Sci.*, **45**, 1229.
40. Isenberg, I., Baird, S. L. and Szent-Györgyi, A. (1961). *Proc. natn. Acad. Sci.*, **47**, 245.
41. Sakurai, T. and Hosoya, H. (1966). *Biochim. biophys. Acta*, **112**, 459.
42. Sarma, R. H., Dannies, P. and Kaplan, N. O. (1968). *Biochemistry*, **7**, 4359.
43. Beinert, H. (1956). *J. Am. chem. Soc.*, **78**, 53523.
44. Holmström, B. (1964). *Photochem. Photobiol.*, **3**, 97.
45. Gibson, Q. H., Massey, V. and Atherton, N. M. (1962). *Biochem. J.*, **85**, 369.
46. Swinehart, J. H. (1965). *J. Am. Chem. Soc.*, **88**, 1056.
47. Strittmatter, P. (1961). *J. biol. chem.*, **236**, 2329 and 2336.
48. Massey, V. and Palmer, G. (1962). *J. biol. Chem.*, **237**, 2347.
49. Yagi, K. and Ozawa, T. (1960). *Biochim. biophys. Acta*, **42**, 381.
50. Tsibris, J. C. M., McCormick, D. B. and Wright, L. D. (1966). *J. biol. Chem.* **241**, 1138.
51. Veeger, C., DerVartian, D. V., Kale, J. F., deKok, A. and Koster, J. F. (1960). "Flavins and Flavoproteins" (Ed. E. C. Slater), p. 242. Elsevier, Amsterdam.
52. Slifkin, M. A. (1963). *Nature*, **197**, 275.
53. Kosower, E. M. (1966). "Flavins and Flavoproteins" (Ed. E. C. Slater). Elsevier, Amsterdam.
54. Karreman, G., Isenberg, I. and Szent-Györgyi, A. (1959). *Science*, **130**, 1191.
55. Slifkin, M. A. and Walmsley, R. H. (1969). *Experientia*, **25**, 930.
56. Rosenthal, I. (1969). *Tetrahed. Lett.* **39**, 3333.
57. Theorell, H. and Nygaard, A. P. (1954). *Acta Chem. Scand.* **8**, 1649; Nygaard, A. P. and Theorell, H. (1955). *Acta chem. scand.* **9**, 1587.
58. Matsunaga, Y. (1966). *Nature*, **211**, 182.
59. Harbury, H. A., LaNue, K. F., Loach, P. A. and Amick, R. M. (1959). *Proc. natn. Acad. Sci.* **45**, 1708.
60. Kolziol, J. and Knobloch, E. (1965). *Biochim. biophys. Acta*, **102**, 289.
61. Bear, C. A., Waters, J. M. and Waters, T. N. (1970). *Chem. Comm.* 702.
62. Pullman, B. and Pullman, A. (1963). "Quantum Biochemistry." Interscience, New York.
63. Langhoff, C. A. and Fritchie, C. J. (1970). *Chem. Comm.* 20.
64. Yagi, K., Okamura, K., Naoi, M., Sugiura, N. and Kotaki, A. (1967). *Biochim. biophys. Acta*, **146**, 77.
65. Yagi, K., Ozawa, T. and Naoi, M. (1969). *Biochim. biophys. Acta*, **185**, 31.
66. Yagi, K., Ozawa, T., Naoi, M. and Kotaki, A. (1968). "Flavins and Flavoproteins" (Ed. K. Yagi), p. 237. University of Tokyo Press, Tokyo.

CHAPTER 8

The Coenzymes

8.1 Introduction

In this chapter the interactions of coenzymes and their analogues other than FMN and ATP will be considered. The interaction of indole containing coenzyme models are dealt with in Section 5.17.

8.2 Pyridinium Salts

8.2.1 ABSORPTION SPECTROSCOPY

A vast body of work has been carried out on coenzyme models by Kosower and his collaborators (1). Various pyridinium salts have been utilized as coenzyme models. It has been well established that solutions of these salts show marked deviations from the Beer-Lambert law (2, 3). The salts exist in solution as complexes between the pyridinium cation and the anion. There is an associated charge transfer band arising from the transfer of an electron from the anion to the π-electron system of the pyridinium ring. The position of this charge transfer band is very solvent sensitive moving to longer wavelengths in solvents of higher polarity (4, 5). The association constants of the

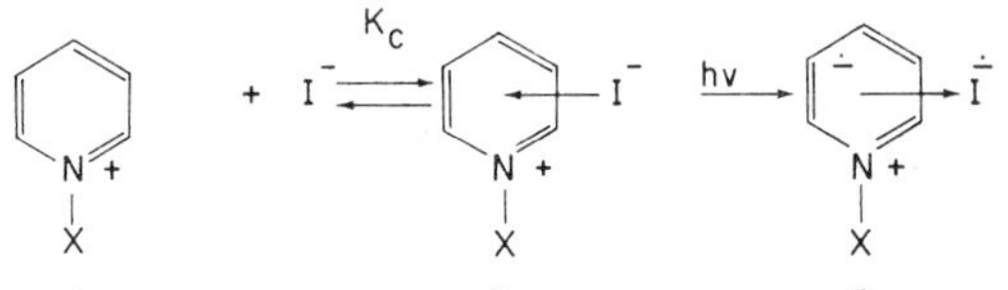

Fig. 8.1. Structures of (1) 1-alkylpyridinium anion; (2) 1-alkyl-pyridinium iodide charge transfer complex; (3) excited state of (2).

complexes evaluated from the spectra, increase with decreasing polarity, or more exactly, with the ion solvating power of the solvent (6). The position of the charge transfer band is also very sensitive to substituents in the pyridinium ring. Electron-donating substituents lower the association constants.

TABLE 8.1

Data for pyridinium ions

Ion	$\lambda_{max}(nm)^a$	$\log \varepsilon^a$	$K_c(M^{-1})^b$
1-methyl	2588	3.68	2.3 ± 0.3
1,2-dimethyl	2654	3.74	
1,4-dimethyl	2553	3.56	
1,2,6-trimethyl	2728	3.89	1.6 ± 0.3
1,2,4,6-tetramethyl	2690	3.88	1.8 ± 0.3
3-carbomyl-1-methyl	2650	3.60	
3-carbomyl-1-benzyl	2650	3.64	

[a] Ref. 4.   [b] Ref. 6.

TABLE 8.2

Solvent effect on the charge transfer maxima of
1-ethyl-4-carbomethoxypyridinium iodide

Solvent	$\lambda_{max}(nm)$	Z (kcal/mole)
methanol	342	83.6
ethanol	359	79.6
2-propanol	375	76.3
1-propanol	365	78.3
butanol	401	71.3
formamide	343	83.3
dimethylformamide	417	68.5
acetonitrile	401	71.3
chloroform	449	63.2
dimethylsulphoxide	402	71.1

Taken from Table 2, Kosower, E. M. (1966). "Flavins and Flavoproteins," (Ed. E. C. Slater). Elsevier, Amsterdam.

The spectra of pyridinium iodide salts also show the spectral bands characteristic of the process $I^- \xrightarrow{h\nu} I \cdot + e$, so establishing that transfer of charge takes place from the anion to the cation (7).

8.2.2 FLASH PHOTOLYSIS

Flash photolysis studies of a pyridinium iodide add confirmation to the charge transfer theory of the pyridinium salts, as absorption of light leads to depletion of the iodine atoms (8).

8.2.3 Complexes of Oxidized and Reduced Coenzyme Models

Other model systems have been studied by Cilento and his coworkers. Cilento and Schrier have demonstrated that complexes can be formed between reduced and oxidized coenzyme models (9). Thus a complex is formed between

TABLE 8.3

Complexes of pyridinium salts and 1,4-dihydropyridines

Salt	Dihydropyridine	λ_{max}(nm)	K_c(M^{-1})
ClBC	ClBCH	440	3.76
NAD	NADH	420	0.3
		440	0
NAD	ClBCH	420	0
ClBC	NADH	410	1.96
ABC	NADH	440	1.62

NAD = nicotinamide adenine dinucleotide. ClBC = 1-benzyl-3-carboxamide-pyridinium chloride. ABC = 1-benzyl-3-acetyl-pyridinium chloride.
In pH 7 buffer at 25°C. Reproduced with permission from Table 1, ref. 9.

1-benzyl-3-carboxamide-pyridinium chloride and 1-benzyl-1,4-dihydronico tin amide. However, no complexing has been observed between the coenzymes themselves in their oxidized and reduced forms. NAD and NADH do not apparently form a complex but this may be due to the pyridinium ring of NAD already being involved in an intramolecular complex with the adenine chromophore. Evidence is presented that a charge transfer complex exists between adenosine or ADP and 1-benzyl-3-carboxamide pyridinium chloride.

TABLE 8.4

Charge transfer complexes of adenosine and ADP with
1-benzyl-3-carboxamide pyridinium chloride

Donor	Temperature (°C)	pH	λ_{max}(nm)	K_c(M^{-1})
adenosine	24.3	7	325	2.34
	43.1	7	325	1.97
ADP	24.2	7.6	330	4.62

Taken from Table 2, ref. 9.

8.2.4 Nuclear Magnetic Resonance

Proton magnetic resonance studies on various NAD analogues also indicate that intramolecular interaction can take place between the adenine and pyridinium rings (10). Vertical stacking of these analogues also takes place.

8.2.5 REDUCED COENZYME MODEL AS DONOR

The charge donor behaviour of a reduced coenzyme model, 1-benzyl-1,4-dihydronicotinamide has been demonstrated as in the presence of chloranil, the spectrum of the chloranil anion appears (11).

8.2.6 AROMATIC HYDROCARBONS

Condensed ring aromatic hydrocarbons form 1 : 1 complexes with 3-carboxamide pyridinium chloride in aqueous ethanol at room temperature (12). The association constants are all very small and no not correlate with

TABLE 8.5

Complexes of hydrocarbons with 1-benzyl-3-carboxamide-pyridinium chloride in aqueous ethanol at room temperature

Hydrocarbon	$\lambda(nm)^a$	$K_c(M^{-1})^a$	$I_p(eV)^b$
naphthalene	326	0.17	8.1
phenanthrene	352	0.76	8.09
1,2,5,6-dibenzanthracene	354	1.03	7.80
pyrene	353	1.09	7.55
1,2-benzanthracene	373	0.27	7.45
anthracene	385	0.13	7.37
3,4-benzopyrene	408	0.67	7.19
3,4-benzphenanthrene	337	0.55	
9-methyl-anthracene	400	0.54	
acenaphthene	330	0.32	
3-methyl-cholanthrene	396	1.35	

Correlation coefficient of λ vs. $I_p = -0.89$ at $\gg$ than 99.5 % confidence level. There is no correlation between K_c and I_p.

[a] Ref. 12.
[b] Slifkin, M. A. (1963). *Nature*, **200,** 877.

hydrocarbon ionization potentials. A linear correlation exists between the ionization potentials and the position of maximum absorption of the charge transfer bands. Increasing the temperature of the solutions causes an increase in some of the association constants but a decrease in others. The authors explain these rather anomalous results as implying that desolvation of the components on complexing occurs thus giving rise to a positive enthalpy of dissociation because of an increase in the entropy.

8.2.6.1 *Halogenated Tyrosines*

Other aromatic hydrocarbons forming complexes with 1-benzyl-3-carbox-amide are the amino acid tyrosine and some halogenated derivatives. Spectrophotometric and thermodynamic data for these complexes have been measured by Cilento and Berenholc (13). The enthalpies of dissociation show a decrease with increasing halogenation. This is the expected results for an interaction

TABLE 8.6

Data of complexes of pyridinium compounds

Pyridinium	Compound	$K_c(M^{-1})$	λ(nm)	ΔH kcal/mole	ΔG	ΔS (e.u.)
ClBC	tyrosine	0.99_{270}	340	-3.38	-0.03	-11.42
ClBC	3-iodotyrosine	1.7				
ClBC	3,5-diiodotyrosine	$2.5_{25.70}$	340	-1.38	-0.52	-2.86
ClBC	3,5-dibromotyrosine	$2.04_{27.20}$		-1.74	-0.42	-4.38
ClBC	3,5-diiodothyronine	0	325			
NAD	tyrosine	0.85_{250}	350			
NAD	3-iodotyrosine	2.73_{250}	335			

All in aqueous acid.
From ref. 13.

involving charge transfer as the halogens are strong electron withdrawing substances. This is the opposite effect to what is observed with tyrosine flavin complexes (Sections 7.3 and 7.8.9), and can perhaps be taken to show that the tyrosines act as π-donors towards pyridinium so that the halogens decrease the overall donor properties of the tyrosines which they do not do in their complexes with the flavin where probably they behave as *n*-electron donors.

8.2.7 INTRAMOLECULAR COMPLEXES

Shifrin has performed a series of studies on the intramolecular complexing of coenzyme models to which are attached various electron donor groups. Salts of 3-carbamoylpyridinium containing different aromatic groups such as indole, phenol, imidazole etc., have been synthesized (14). In all cases new bands are observed arising from the intramolecular charge transfer interaction. The positions of maximum absorbance of these bands are linearly dependent on the energy of the highest occupied molecular orbital of the donor. The charge transfer absorption in the salt of imidazole dissappears on acidification which is interpreted as meaning that the charge transfer is due to the lone-pair electrons on the nitrogen in the indole ring which become blocked by protonation in acid media.

TABLE 8.7

Absorption properties of β-substituted 1-ethyl-3-carbamoylpyridinium chlorides

R	λ_{max}(nm)	Energy of highest filled orbital	
indole	325	0.52	
thiomethyl	300		
phenol	296	0.73	
imidazole	294	0.65	
benzene	283	1	

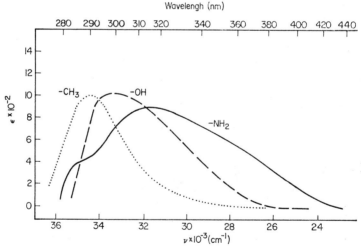

Coefficient of correlation $= -0.86$ at 96% confidence level.
Reproduced with permission from Table 2, ref. 14.

The addition of different groups to the phenyl derivative of the pyridinium salt has been studied also (15). The wavelength of maximum absorption of the intramolecular charge transfer complex shows a very linear correlation

Fig. 8.2. Intramolecular charge transfer spectra of *para*-substituted phenylethylnicotin-amides in which the substituent is amino- (———), hydroxy- (—–—–) and methyl- (. . . . .). Reproduced with permission from Fig. 2, ref. 15.

with the ionization potentials of the same groups substituted into benzene. The ionization potentials increase in the order

$$-NH_2 < -OCH_3 < -OH < -CH_3 < -Cl < -H.$$

A purine substituted pyridinium salt has been synthesized and there is some indication that this model compound also exhibits a charge transfer band although this band is too close to existing bands of the unsubstituted salt to be characterized (16).

TABLE 8.8

Absorption properties of substituted
N-(β-p-x-phenylethyl)-3-carbamoylpyridinium halides

Substituent	$\lambda_{max}(nm)^a$	I_p (subst. benzene) (eV)^b	$\varepsilon_c{}^a$
— NH$_2$	315	7.7	910
— OH	298	8.5	1040
— OCH$_3$	295	8.2	1160
— CH$_3$	290	8.8	1010
— Cl	286	9.1	950
— H	283	9.2	1080

Correlation coefficient of λ vs. $I_p = -0.95$ at $\gg 99.5\%$ confidence level.
^{a} From Table 2, ref. 16.
^{b} Watanabe, K., Nakayama, T. and Mottl, J., (1962). *J. Quant. Spect. Rad. Transfer*, **2**, 369.

N-(β-p-x-phenylethyl)-3-carbamoylpyridinium iodide.

8.3 Methyl Nicotinamide

8.3.1. ASCORBIC ACID

Solutions of nicotinamide and ascorbic acid are yellow (17). This colouration is dependent on the temperature, decreasing with increasing temperature. Spectral studies show that the absorbance of the mixed solutions obeys the Benesi-Hildebrand equation. Moreover, a Job plot confirms the 1 : 1 nature of the complex. The absorbance at a given wavelength is a function of the pH and peaks at *ca.* pH 4 which lies midway between the pK$_a$ values of protonated nicotinamide and ascorbic acid. Ascorbic acid also interacts with 3-carbamoyl-1-methyl-pyridinium chloride and has a maximum absorbance at pH 6. As the ionization of N'-methyl nicotinamide is pH independent, this implies that the observed interaction is between the ascorbic acid cation and the nicotinamide protonated base. The enthalpy of dissociation is -1.5 kcal/mole.

8.3.2 CHLOROTHEOPHYLLINATE

The complexing of N'-methyl nicotinamide with the electron donor 8-chlorotheophyllinate anion is very dependent of the nature of the salts used to control the ionic strength (18). This is because of competitive complexing between the nicotinamide and the various cations. In the presence

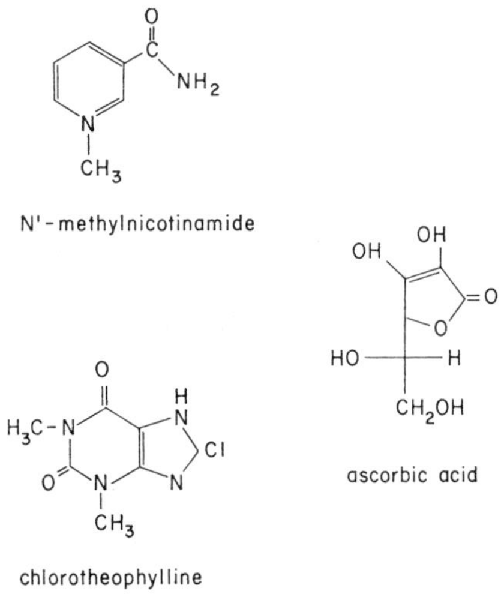

N'-methylnicotinamide

ascorbic acid

chlorotheophylline

Fig. 8.3.

of sodium perchlorate, the apparent association constant of the nicotinamide and theophylline anion is 1.52 M^{-1}, and in the presence of sodium acetate, 2.59 M^{-1}. Taking into account the competitive complexing with the perchlorate using a variant of the Benesi-Hildebrand equation for the formation of two complexes simultaneously in solution (Section 1.7) it is found that the true association constant of the theophyllinate complex is 2.67 M^{-1} and that of the perchlorate complex is 0.86 M^{-1}.

8.3.3 CHLOROTHEOPHYLLINATE AND COENZYME MODEL

The hydrolysis of the ester 3-carbo-methoxy-1-methylpyridinium is hindered by 8-chlorotheophyllinate anion (19). The anion forms a complex with pyridinium. Application of the Benesi-Hildebrand equation to spectra of mixtures of these two compounds yields association constants of 4.7 M^{-1} at an ionic strength of 0.2 and 3.9 M^{-1} at an ionic strength of 0.3. Charge

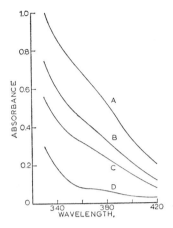

Fig. 8.4. CT = Chlorotheophyllinate; NMN = N'-methyl nicotinamide. Plot illustrating the influence of added salts on the spectral characteristics of solutions at a pH of 8.0 containing 0.1 M CT and 0.005 M NMN. Curve A represents the spectrum of the mixture in the absence of added salt. Curve B is the spectrum of the mixture in the presence of 0.9 M sodium acetate. Curve C is the spectrum of the mixture in the presence of 0.9 M sodium perchlorate. Curve D is the spectrum of 0.1 M CT alone. Reproduced with permission from Fig. 1, ref. 18.

donation from the anion to the pyridinium ring reduced the electron withdrawal ability of the ring, resulting in a lowering of acyl activation of the ester carbonyl. Additionally the presence of the anion in the complex with the ester may cause steric hindrance preventing the approach of the nucleophilic reagents.

8.4 NAD$^+$

NAD$^+$ like protonated nicotinamide gives strongly coloured solutions with a variety of electron donors including some pteridines, indoles, uric acid and some phenothiazines (20). The positions of the maxima of the bands giving rise to the colour correlate with the ionization potentials of the electron donors and are thus charge transfer bands (see Table 8.9).

8.5 Alkylpyridinium Iodides

The interaction between miscelles of long-chain alkylpyridinium iodides and various common anions in aqueous solution of chloroform has been studied by Ray and Mukerjee (21, 22, 23).

TABLE 8.9

Charge transfer absorptions of NAD$^+$ complexes

Donor	λ_{max}(nm)	I_p (eV)	
2-amino-4,7-dihydroxy-6-methylpteridine	415		a
2-amino-4-hydroxy-6,7-dimethylpteridine	423		a
serotonin creatine sulphate	320		b
lysergic acid	340	7.8	b
indole	310	7.9	b
uric acid	340	7.5	b
promazine hydrochloride	380	7.2	b
promethazine hydrochloride	385	7.2	b
methdilazine hydrochloride	380	7.2	b
chlorpromazine hydrochloride	390	7.3	b

Correlation coefficient $= -0.95$ at $\gg 99.5\%$ confidence level.
[a] In 5N HCl.
[b] pH 7 aqueous buffer.
From Table 1, ref. 20.

In the presence of some of these anions in chloroform charge transfer bands appear which can be easily recognized as such, as the positions of the maxima are in the order of the electronegativity of the ions (21). In aqueous solution, many of the anions interact with pyridinium but not necessarily the same ones as in chloroform. The extinction coefficients of the solution show a marked dependence on the solvent, the values being much lower in chloroform than in the more polar alcohols. In the more polar solvents the complexes are less dissociated than in chloroform. Geometrical effects are also important in these interactions as no charge transfer is detected between either I^- or Br^- with tetraalkylammonium or trialkylphenylammonium ion, suggesting that the planar geometry of the nitrogen charge centre in the alkylpyridinium ions facilitates charge transfer with the ions (22).

TABLE 8.10

Enthalpies of dissociation of some dodecyl colloids

Colloid	ΔH (kcal/equiv.)	
sodium dodecyl sulphate	$0.4_{20°}$	$-1.1_{40°}$
dodecylsulphonic acid	$0.5_{20°}$	$-1.1_{40°}$
dodecylammonium chloride	$1.1_{20°}$	$-2.0_{50°}$
dodecylpyridinium bromide	$-0.55_{20°}$	$-1.77_{40°}$
dodecylpyridinium iodide	$-1.3_{21°}$	$-2.6_{40°}$

Reproduced with permission from Table 2, ref. 23.

The adsorption of iodide and chloride onto micelles of another coenzyme model, dodecylpyridinium has been studied and the enthalpies of adsorption measures (23). It is presumed that charge transfer forces play a major role in binding the ions to the micelles.

8.6 Lipoic Acid (Thiotic Acid)

8.6.1 TETRACYANOETHYLENE COMPLEX

Lipoic acid, a coenzyme for the oxidative decarboxylation of pyruvic acid forms 1 : 1 complexes with tetracyanoethylene in dichloromethane together with other disulphides (24). These complexes are characterized by charge

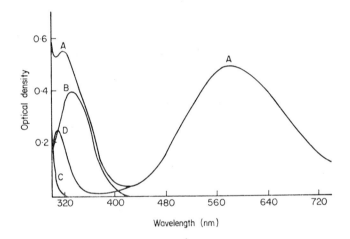

dl-thiotic acid

Fig. 8.5 (upper). Resolution of spectrum of dichloromethane solution, 2.786×10^{-2} M in *dl*-thioctic acid and 4.726×10^{-2} M in TCNE. A, Total absorption; B, free $-$ S $-$ S $-$; C, free TCNE; D, complexed $-$ S $-$ S $-$. Reproduced with permission from Fig. 1, ref. 24.

transfer bands in the visible. The band of the complexed disulphide shows a short wavelength shift compared to the disulphide band of the free molecule.

The enthalpy of dissociation of the lipoic acid tetracyanoethylene complex is -7.2 kcal/mole. The ionization potential estimated from the charge transfer band is 7.53 eV.

8.6.2 FLAVIN MONONUCLEOTIDE

Lipoic acid in common with other electron donors (25), *cf.* Section 7.2, has a marked inhibition on the photo reduction of FMN by NADH. Lipoic acid has a strong quenching effect on the fluorescence of FMN (26).

REFERENCES

1. Kosower, E. M. and Skorcz, J. A. (1962). "Advances in Molecular Spectroscopy," p. 413. Academic Press, New York; and references cited therein.
2. Hantzsch, A. (1911). *Ber.* **44**, 1783.
3. Kosower, E. M. (1955). *J. Am. chem. Soc.* **77**, 3383.
4. Kosower, E. M. and Klenedinst, P. E. (1956). *J. Am. chem. Soc.* **78**, 3493.
5. Kosower, E. M. (1956). *J. Am. chem. Soc.* **78**, 5700.
6. Kosower, E. M. and Burbach, J. C. (1956). *J. Am. chem. Soc.* **78**, 5853.
7. Franck, J. and Scheibe, G. (1928). *Z. phys. Chem.* **A139**, 22.
8. Kosower, E. M. and Lindquist, L. (1965). *Tetrahed, Lett.* **50**, 4481.
9. Cilento, G. and Schrier, S. (1964). *Archs Biochem. Biophys.* **107**, 102.
10. Sarma, R. H., Dannies, P. and Kaplan, N. O. (1968). *Biochem.* **7**, 4359.
11. Cilento, G. and Zinner, K. (1966). *Biochim. biophys. Acta*, **120**, 84.
12. Cilento, G. and Sanioto, D. L. (1965). *Archs Biochem. Biophys.* **110**, 133.
13. Cilento, G. and Berenholc, M. (1965). *Biochim. biophys. Acta*, **94**, 271.
14. Shifrin, S. (1964). *Biochemistry.* **3**, 829.
15. Shifrin, S. (1965). *Biochim. biophys. Acta*, **96**, 173.
16. Shifrin, S. (1968). "Molecular Associations in Biology" (Ed. B. Pullman), p. 339. Academic Press, New York.
17. Guttman, D. E. and Brooke, D. (1963). *J. Pharm. Sci.* **52**, 941.
18. Brooke, D. and Guttman, D. E. (1968). *J. Pharm. Sci.* **57**, 1206.
19. Brooke, D. and Guttman, D. E. (1968). *J. Pharm. Sci.* **57**, 1677.
20. Fulton, A. and Lyons, L. E. (1967). *Aust. J. Chem.* **20**, 2267.
21. Ray, A. and Mukerjee, P. (1966). *J. phys. Chem.* **70**, 2138.
22. Mukerjee, P. and Ray, A. (1966). *J. phys. Chem.* **70**, 2144.
23. Mukerjee, P. and Ray, A. (1966). *J. phys. Chem.* **70**, 2150.
24. Moreau, W. M. and Weiss, K. (1965). *Nature*, **208**, 1203.
25. Radda, G. K. and Calvin, M. (1963). *Nature*, **200**, 464.
26. Radda, G. K. (1966). *Biochim. biophys. Acta*, **112**, 448.

CHAPTER 9

Miscellaneous Molecules

9.1 Steroids and Sterols

9.1.1 INTERACTION WITH ELECTRON ACCEPTORS

The steroids have been shown to give charge transfer bands on mixing with electron acceptors such as chloranil, although in many cases these bands are poorly resolved (1, 2). Polarographic measurements have been carried out on some natural and synthetic oestrogens to obtain estimates of their ionization potentials and electron affinities in order to see whether the coloured bands

TABLE 9.1

First half-wave potentials of some hormones. Comparable solvation energy terms are involved in each of the groups. Donor and acceptor strengths increase as $E_{\frac{1}{2}}^{oz}$ decreases and $E_{\frac{1}{2}}^{red}$ becomes less negative

Compound	V $vs.$ S.C.E.	
	$E_{\frac{1}{2}}^{ox}$	$E_{\frac{1}{2}}^{red}$
stilboestrol	0.90	-2.80
equilenin	1.10	-2.17
oestrone	1.19	-1.98
β-oestradiol	1.24	N
progesterone	> 2.2	-2.27
11-dehydroprogesterone	> 2.2	-2.24
11-ketoprogesterone	> 2.2	-2.22
testosterone	> 2.2	-2.26
androstane-17β-ol-3-one	$ca.$ 2.3	-2.82
androstanedione	> 2.2	-2.25
$\Delta'E$ prednisone	2.24	-1.88
cortisone	> 2.2	-2.12
hydrocortisone	> 2	-2.18
dihydrocortisone	2.04	-2.43
prolactin	0.94	

Reproduced with permission from Table 1, ref. 1.

obtained on mixing with electron acceptors do correlate with one or other of these parameters (1). Plots of the oxidative potential *vs.* the position of maximum absorption of the charge transfer bands with chloranil in methyl cyanide do show a linear correlation. These oestrogens have large electron affinities, as deduced from their reductive potentials and might therefore be expected to behave also as electron acceptors in the presence of donors. They also have an affinity for thermal electrons (58). Szent-Györgyi believes that the interaction of these steroids with iodine in water, where iodine is present as I_3^- is evidence of the acceptor role of the steroids (3). He suggests that a complex is formed between the steroid and the triiodide ion as donor. However, several other authors have shown that charge donors interact with iodine in water to produce the triiodide ion as a byproduct (4).

9.1.2 IODINE

Wobschall and Norton have also carried out work on steroid iodine complexes (5). 1 : 1 complexes are formed in freon solution at low concentration. These complexes are characterized by new absorption bands in the near ultraviolet and rather low values of association constants. These complexes are formed with different functional groups of the steroids and are of both

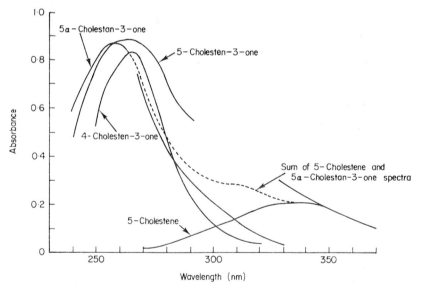

Fig. 9.1. Charge transfer spectra of iodine complexes of selected steroids. Note that the sum of the spectra of the 5-cholestene complex and the 5α-cholestan-3-one complex is qualitatively the same as that of the 5-cholesten-3-one spectrum but different from the 4-cholesten-3-one spectrum. Reproduced with permission from Fig. 1, ref. 5.

TABLE 9.2

Spectral maxima and thermodynamic constants in freon for 1 : 1 iodine
charge transfer complexes with cyclohexane derivatives

Compound	λ_{max} (mμ)	Log ε	K_1 (20°) (liters/ mole)	$-\Delta H$ (kcal/ mole)	Type complex	Foot- note	Ref.
A Cyclohexane	243 ± 5	(4.28)	(0.013)	—	σ, σ	a	e
B Cyclohexene	300 ± 10	4.15	0.35	2.2	π, σ	b, c	f
C Cyclohexanol	233 ± 5	4.02	2.2	—	n, σ		g
D Cyclohexanone	253 ± 5	3.99	2.4	4.8	$n + \pi, \sigma$	d	g
E 2-Cyclohexenone	249 ± 5	4.25	3.2	4.1	$n + \pi, \sigma$		g

a Probably a contact complex, so that K_1 is not well defined.
b Solvent is CCl_4.
c Iodinates rapidly.
d Forms dimers.
e Evans, D. F. (1955). *J. chem. Phys.* **23**, 1424.
f Trayham, J. G. and Olechowski, J. R. (1959). *J. Am. chem. Soc.* **81**, 471.
g Wobschall, D. and Norton, D. A. (1965). *J. Am. chem. Soc.* **87**, 3559.
Reproduced with permission from Table 1, ref. 5.

n and π-type. Steroids having more than one functional group can form more
than one type of complex simultaneously and they then possess multiple
charge transfer bands. At higher concentration, additional bands appear.
These bands are attributed to the formation of steroid iodine dimers, but as
pointed out by the authors the bands are very close to those of the triiodide
ion. One well-known feature of complexes involving iodine especially in
polar solvents is the appearance of the triiodide ion as a byproduct (examples
are given in Sections 3.8 and 5.3). The reason that the authors assign these
new bands to the dimer complex is that the absorbance is proportional to the
square of the monomer concentration, according to the relationship

$$[S..I_2]_2 \rightleftharpoons K_2[S..I_2]^2 \rightleftharpoons K_1 K_2[S]^2[I_2]^2,$$

where S is the steroid, K_1 and K_2 are the association constants for the monomer
and dimer respectively and $S..I_2$ is the complex.

This dimer complex then can transform to give the triiodide ion spectrum

$$S..I..I..S \rightleftharpoons S..I^+..I_3..S.$$

This is a different mechanism from that usually advanced to explain the
appearance of the triiodide ion in systems where complexing occurs *viz.*:

$$A + 2I_2 \rightleftharpoons (AI)^+..I_3^-.$$

The spectra of solid steroid iodine complexes are very similar to those of

TABLE 9.3

Spectral maxima and equilibrium constants in freon for 1 : 1 iodine charge transfer complexes with monofunctional steroids[a]

Steroid	λ_{max}(nm)	Similar compound Table 9.2	K_1 (20°) (M^{-1})	Type complex
5 α-Cholestane	~ 240	A	(0.06)	σ, σ
5-Cholestene	335 ± 20	B	1.0	π, σ
5 α-Cholestan-3β-ol	230 ± 5	C	2.1	n, σ
5 α-Cholestan-3α-ol	231 ± 5	C	1.3	n, σ
5 α-Cholestan-3-one	257 ± 3	D	2.0	$n + \pi, \sigma$
4-Cholesten-3-one	266 ± 3	E	13.0	$n + \pi, \sigma$
5-Cholesten-3β-ol[b]	320 ± 20	B	0.4	π, σ
3,5-Cholestadiene[b, c]	370 ± 20	—	1.3	π, σ
5 β-Cholestan-3-one	250 ± 3	D	2.6	$n + \pi, \sigma$

[a] A carbon-carbon double bond will be considered here as a functional group.
[b] Not recrystallized.
[c] Additional strong absorption at 280 nm forms rapidly. Black precipitate forms on standing if kept cool and dark; otherwise iodination occurs.

5 α-cholestan-3β-ol (cholestanol)

Reproduced with permission from Table 1, ref. 5.

the dimer. A blue complex can be precipitated from ethanol-water solutions whose spectrum has a broad band in the visible as well as the typical triiodide bands in the ultra-violet. Water is essential for the blue colour.

These blue complexes bear a great similarity to the well-known amylose-iodine complexes (6), used as a test for the detection of iodine. These amylose complexes are not thought to be of the charge transfer type but to involve covalent bonding of a metallic nature.

Jones et al. (7) have produced brown solids presumed to be charge transfer complexes on evaporating solutions of oestrogens and iodine (I_2) in chloroform. No ESR signals are cahibited by these complexes.

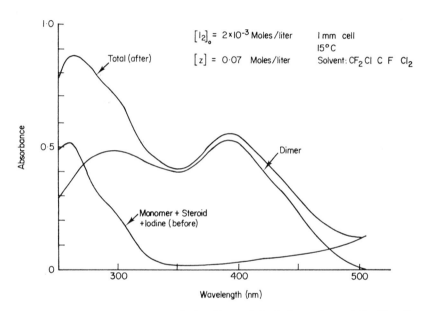

Fig. 9.2. Absorption spectrum of the dimer of 5α-cholestan-3-one + I_2. The dimer absorption spectrum was obtained by measuring the change in absorption before and after the slow formation of the dimer in solution. Reproduced with permission from Fig. 2, ref. 5.

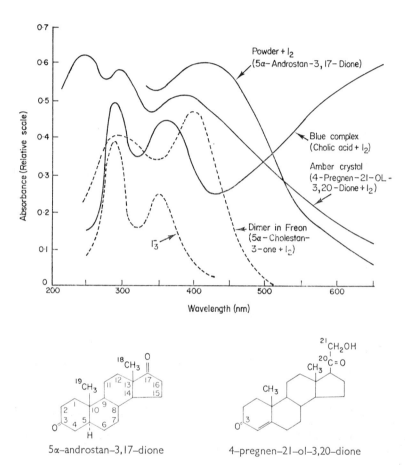

Fig. 9.3. Absorption spectra of solid steroid-iodine complexes. The sample labelled "powder + I_2" was prepared by exposing a film of steroid (sublimed under vacuum) on a quartz plate to iodine vapour while in the spectrophotometer. The "cholic acid blue complex" was precipitated from water-ethanol solution and suspended in water. The "amber crystal" sample was obtained by smearing crystals onto a quartz plate. The I_3^- spectrum is produced by a mixture of KI and I_2 in water. The "dimer in solution" sample is the same as shown in Fig. 9.2. Reproduced with permission from Fig. 3, ref. 5.

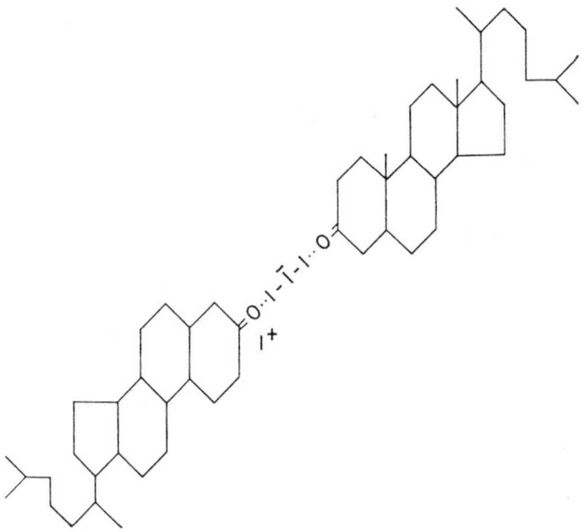

Fig. 9.4. Postulated dimer structure. Reproduced with permission from Fig. 4, ref. 5.

TABLE 9.4

Measured thermodynamic constants for monomer and dimer
steroid-iodine charge transfer complexes

Compound	$K_1(20°)$ (M^{-1})	$-\Delta G^a$ (kcal/mole)	$-\Delta H$ (kcal/mole)	$-S^a$ (e.u.)
Monomers				
5 a-Cholestan-3-one	2.9	0.6	4.3	10
4-Cholesten-3-one	9.0	1.3	5.8	15
5 a-Androstan-3,20-dione	2.5	0.5	3.7	8
Probable error	$\pm 30\%$	± 0.3	± 0.5	± 2
	$K_2(20°)$ (M^{-1})	$-\Delta G_2{}^a$ (kcal/mole)	$-\Delta H_2$ (kcal/mole)	$-\Delta S_2{}^a$ (e.u.)
Dimers				
5 a-Cholestan-3-one	350	3.4	15	41
4-Cholesten-3-one	100+	—	—	—
5 a-Androstan-3,20-dione	300	3.3	16	43
Probable error	$+ 100\%$ $- 50\%$	± 0.5	± 4	± 10

[a] When K is expressed in terms of liters/mole.
[b] Order-of-magnitude value-rate of iodination was comparable to the rate of dimerization.

TABLE 9.4—*cont.*

Spectral maxima and equilibrium constants in freon for iodine charge transfer complexes with steroids having more than one functional Group[a]

| Steroid | First Peak | | | |
	λ_{max} (nm)	Similar compound Table 9.2	$K_1 (20°)$ (M^{-1})	Footnote
5-Cholesten-3-one	310 ± 20	B	0.6	
5-Cholesten-3 β-ol	330 ± 20	B	0.6	c
5a-Androsten-3,20-dione	258	D	2.9	d
4-Pregnen-3,20-dione	264	E	10	d
4-Pregnen-21-ol-3,20-dione	263	E	9	c
4-Androsten-17β-ol-3-one	265	E	12	c
5a-Androstan-3β-ol-17-one	265	D	2.6	

| Steroid | Second Peak | | | |
	λ_{max}[b]	Similar compound Table 9.2	$K_1 (20°)$ (M^{-1})	Footnote
5-Cholesten-3-one	260	D	1.7	
5-Cholesten-3β-ol	(253)	C	—	c
5a-Androsten-3,20-dione	(257)	D	—	d
4-Pregnen-3,20-dione	(253)	D	—	d
4-Pregnen-21-ol-3,20-dione	(253, 235)	D, C	—	c
4-Androsten-17β-ol-3-one	(235)	C	—	c
5a-Androstan-3β-ol-17-one	235	C	2	

[a] A carbon-carbon double bond is considered here as a functional group.

[b] Values in parentheses are expected but not observed (see footnotes c and d).

[c] The OH complex peak occurs below the wavelength at which the steroid becomes opaque (250 nm), and thus was not observed.

[d] Both maxima are expected to occur at the same wavelength and have the same equilibrium constants.

Reproduced with permission from Table 2, ref. 6.

TABLE 9.5

Equilibrium constants and spectrophotometric data for donor-acceptor
complexes of lipids with 2,4-dinitrophenol in CCl_4 at 24°C.

Conc. of 2,4-DNP (acceptor) (mol/1) $\times 10^5$	Lipids (donor)	Conc. range of lipids mole/1 $\times 10^3$	Equil. constant, K_c (M^{-1})	Donor-acceptor band	
				λ_{max} (nm)	ε_{max} $\times 10^4$
4.25	egg lecithin	0.69–2.06	720.0	310	1.265
8.50	synthetic lecithin	2.45–7.35	49.3	300	1.138
17.00	oxidized cholesterol	15.10–75.50	1.9	291	1.010

Reproduced with permission from Table 1, ref. 9.

9.1.3 CHOLESTEROL

9.1.3.1 *Semiconductivity of Cholesterol Complexes*

The semiconductivity of cholesterol is markedly affected by the presence
of organic electron acceptors whether in powder form or as a bimolecular
lipid film (8, 9). The effect of adding electron acceptors is to lower the activation energy and to cause spectacular decreases in the resistivity. Such
behaviour in organic systems can be caused by the formation of charge
transfer complexes (Section 2.4). The effect of the acceptors is in the order
I_2 > picric acid > dinitrophenol > trinitrobenzene.

9.1.3.2 *Iodine*

The absorption spectrum of oxidized cholesterol with iodine in carbon
tetrachloride shows a shift of the iodine band from 518 nm in the free state to
509 nm in the complexed state. The association constant of the cholesterol
iodine complex is 6.26 M^{-1} (9).

9.1.4 PURINES, PYRIMIDINES, PORPHYRINS

The interactions with the above named compounds are discussed in Sections
4.10 and 6.8. Although it has been suggested (3) that the steroids should
have electron acceptor character, this has only been found to be so with
purines and pyrimidines (Section 4.10).

9.2 Lecithins

Spectroscopic studies of the interaction of lecithins with iodine in both
polar and non-polar solvents have been made by Rosenberg and coworkers
(10). Lecithin forms a 2 : 1 complex with iodine in carbon tetrachloride. This
is accompanied by the appearance of the triiodide absorption spectrum. The

TABLE 9.6

Semiconductivity data of oxidized cholesterol in solid state (according to $\sigma(T) = \sigma_{0'} \, e^{E/2kT_0} \, e^{-E/2kT}$), $1/2kT_0 = 11.80 \text{ eV}^{-1}$,

$\sigma_{0'} = 4.44 \times 10^{-5} \, (\Omega^{-1} \text{cm}^{-1})$,

State	$E(\text{eV})$	$\sigma_{25°C} \, (\Omega^{-1}\text{cm}^{-1})$	$\sigma_0' e^{E/2kT_0} \, (\Omega^{-1}\text{cm}^{-1})^a$ (calculated)	$\sigma_{0'} \, e^{E/2kT_0} \, e^{-E/2kT} \, (\Omega^{-1}\text{cm}^{-1})$ (measured)
Dry	3.97	1.33×10^{-19}	1.14×10^{16}	1.05×10^{14}
Partially Hydrated	2.99	2.89×10^{-15}	7.05×10^{10}	5.65×10^{10}
Iodine	2.68	1.40×10^{-13}	2.39×10^{9}	5.95×10^{9}
Fully Hydrated	2.15	3.33×10^{-12}	4.77×10^{6}	4.71×10^{6}
Fully Hydrated + 2,4-DNP	1.22	2.44×10^{-10}	7.95×10^{1}	4.74
Fully Hydrated + Iodine	0.81	1.67×10^{-7}	0.63	1.18

Semiconductivity data of bimolecular lipid membranes of oxidized cholesterol (according to $\sigma(T) = \sigma_{0'} \, e^{E/2kT_0} \, e^{-E/2kT}$), $1/2kT_0 = 11.80 \text{ eV}^{-1}$,

$\sigma_{0'} = 1.0 \times 10^{-8} \, (\Omega^{-1}\text{cm}^{-1})$,

Acceptor in aqueous solution with conc.	$E(\text{eV})$	$\sigma_{25°C} \, (\Omega^{-1}\text{cm}^{-1})$	$\sigma_{0'} \, e^{E/2kT_0} \, (\Omega^{-1}\text{cm}^{-1})^a$ (calculated)	$\sigma_{0'} \, e^{E/2kT_0} \, e^{-E/2kT} \, (\Omega^{-1}\text{cm}^{-1})$ (measured)
0	1.98	3.50×10^{-15}	1.45×10^{2}	1.52×10^{2}
Trinitrobenzene (1×10^{-3} M)	1.69	2.18×10^{14}	4.61	4.16
2,4-dinitrophenol (2.33×10^{-4} M)	1.40	2.20×10^{-13}	1.495×10^{-1}	1.26×10^{-1}
2,4-dinitrophenol (4.67×10^{-4} M)	1.30	6.00×10^{-13}	4.63×10^{-2}	5.88×10^{-2}
2,4-dinitrophenol (7.00×10^{-4} M)	1.24	9.00×10^{-13}	2.28×10^{-2}	2.84×10^{-2}
2,4-dinitrophenol (9.33×10^{-4} M)	1.20	1.33×10^{-12}	1.41×10^{-2}	1.74×10^{-2}
Picric acid (1×10^{-3} M)	1.10	2.60×10^{-12}	4.42×10^{-3}	5.28×10^{-3}
Iodine (1×10^{-3} M)	0.80	1.95×10^{-11}	1.27×10^{-4}	1.07×10^{-4}

^{a} These are calculated by using the values for $1/2kT_0$ and $\sigma_{0'}$.
Reproduced with permission from Tables 2 and 3, ref. 9.

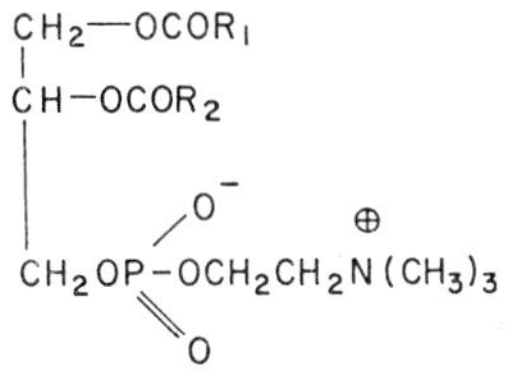

$$CH_2\!-\!OCOR_1$$
$$CH\!-\!OCOR_2$$
$$CH_2OP\!-\!OCH_2CH_2N(CH_3)_3$$

Fig. 9.5. Lecithin. R_1 and R_2 are fatty acid residues.

suggested mechanism is that put forward for other biochemical charge donors *viz.*:

$$(\text{lecithin})_2 + I_2 \rightleftharpoons \text{lecithin} . . I^+ . \text{lecithin} . I^-$$
$$\rightleftharpoons \text{lecithin} . . I^+ + I^- + \text{lecithin}, \ I_2 + I^- \rightleftharpoons I_3^-.$$

In water, the iodine already exhibits the spectrum of the triiodide ion and the addition of lecithin causes an increase in the spectrum of the triiodide ion and a decrease in the spectrum of the solvated molecular iodine (8) in an identical manner to the changes caused by amino acids (11). The spectrum of egg lecithin exposed to iodine vapour again shows the spectrum of the triiodide ion. The spectral changes are very much larger for egg lecithin than for synthetic lecithin (8) showing that the natural product is a much stronger donor.

Egg licithin also causes changes in the absorption spectra of other charge acceptors in carbon tetrachloride (9). Equilibrium constants have been obtained for the charge transfer complexes formed and the order of complexing is iodine > picric acid > carbonyl-cyanide-chlorophenylhydrazone (ClCCP) > dinitrophenol > trinitrobenzene, the identical order observed for complexes of cholesterol (5).

9.3 β-Carotene

9.3.1 SPECTRA OF IODINE CAROTENE MIXTURES

Spectra of mixtures of carotene and iodine have several distinctive features (11, 13). The normal absorption band of β-carotene at *ca.* 460 nm is severely reduced and the spectrum of the triiodide ion is present. These spectra are concentration dependent. In solutions of high dielectric constants a new band occurs at *ca.* 1000 nm. There are also several isosbestic points in the spectra.

9.3.2 SOLID COMPLEXES

β-Carotene displays a strong ESR signal on mixing with iodine (14). This mixture behaves also as a semiconductor with a low activation energy and high conductivity typical of charge transfer complexes. The temperature

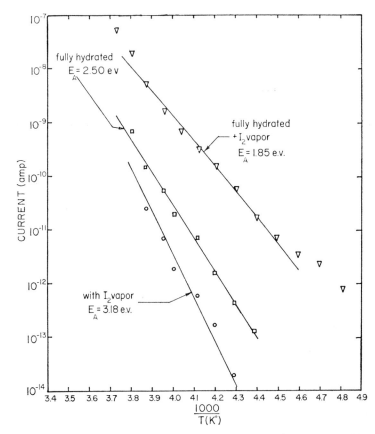

Fig. 9.6. Semiconductive behaviour of an egg lecithin film for various ambient atmospheres. σ_{24} (fully hydrated) $= 2.5 \times 10^{-7} \ \Omega^{-1} \ cm^{-1}$. Reproduced with permission from Fig. 1, ref. 10.

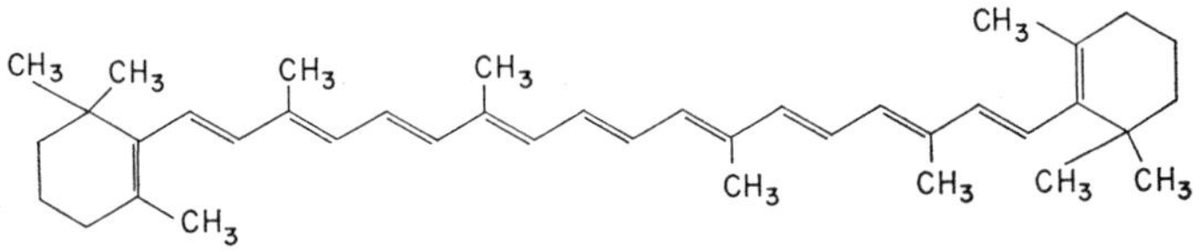

Fig. 9.7. β-Carotene.

dependence of the conductivity is quite different to that of the ESR signal which means that either the ESR intensity is not a measure of the number of conduction carriers or that the determining factor in the temperature dependence of the conductivity is the mobility of the carriers rather than their number (14).

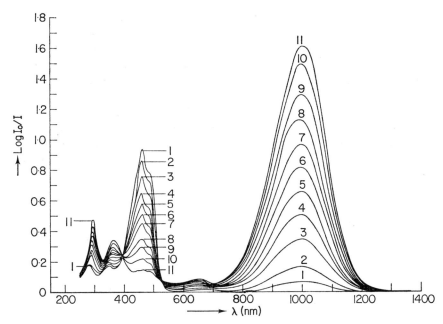

Fig. 9.8. Absorption spectra of 8×10^{-6} M β-carotene in $ClCH_2CH_2Cl$ to which the following amounts of iodine were added: (1) 1×10^{-6} M, (2) 2×10^{-6} M, (3) 4×10^{-6} M, (4) 6×10^{-6} M, (5) 8×10^{-6} M, (6) 1×10^{-5} M, (7) 1.2×10^{-5} M, (8) 1.5×10^{-5} M, (9) 1.8×10^{-5} M, (10) 2.5×10^{-5} M, and (11) 4×10^{-5} M. Reproduced with permission from Fig. 4, ref. 13.

9.3.3 MECHANISM OF INTERACTION

The following reaction scheme explains these results:

$$\text{carotene} + 2I_2 \rightleftharpoons (\text{carotene} .. I)^+ + I_3^-,$$

the carotene forming a charge transfer complex with the I^+ ion. The band at 1000 nm is thought to arise from the charge transfer transition between the complex in its ground state carotene $.. I^+$ and the excited complex carotene$^+ .. I$.

9.3.4 NATURE OF THE NEW ABSORPTION BAND

Ebrey whilst accepting that the absorption band at 1000 nm arises from the charge transfer complex, believes that it is an intrinsic band of carotene shifted to longer wavelength (15). The formation of the complex gives contributions to the structure from ionic forms. A very similar explanation to that put forward to explain the long wavelength shift in chloranil on forming *n*-π complexes with amino acids (Sections 1.12 and 3.3).

9.3.5 SEMICONDUCTIVITY

Eley and Snart (16) have examined the conductivity of films of carotene plus bovine serum albumin and the same components plus chlorophyll and β-methyl-naphthaquinone. The effect of adding carotene to the protein is to lower both the resistivity and the activation energy in a similar manner to both chlorophyll and the quinone (Section 6.10.3). Presumably in this case carotene is functioning as an electron acceptor.

Semiconductivity studies on the effect of adsorbed gases on β-carotene films show that the effect of gases which are electron donors is to markedly decrease the resistivity of the carotene film (17). This decrease mirrors the increasing donor ability of the gases. There are also shifts and changes in the absorption spectra of the films which are dependent on the gases used. New bands seen in the film are not thought to be charge transfer bands. The near infra-red was not however examined.

In this case also carotene is believed to be forming charge transfer complexes acting as an acceptor. Similar effects are observed in films of other carotenes.

9.3.6 CONCLUSION

Carotene is therefore one of at present the few known classes of biomolecules which can function as donors and acceptors under moderate conditions.

9.4 Vitamin A

Studies to ascertain the electron donor and acceptor properties of retinol (vitamin A) have been carried out by Lucy *et al.*

9.4.1 ORGANIC ACCEPTORS

Retinol tetracyamoquinodimethane (TCNQ) mixtures show no sign of interaction in non-polar solvents (18). In polar solvents, the typical spectrum of the TCNQ anion appears (18, 19). There is however no accompanying

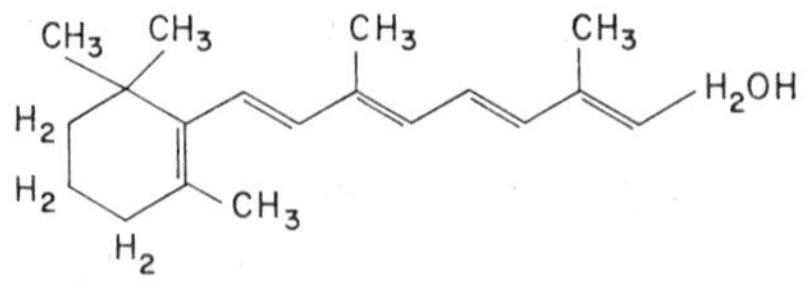

Fig. 9.9. Vitamin A (retinol).

retinol cation. This is analogous to the behaviour of the TCNQ tetramethyl-phenylenediamine complex (20). Retinoic acid behaves similarly to retinol.

Mixed solutions of retinol and chloranil in dimethyl formamide are green the colour arising from a new absorption band at 607 nm. This might be a charge transfer band. Increasing the polarity of these solutions by adding phosphate buffer causes the appearance of a new sharp band at 400 nm. The addition of sodium chloride to these solutions causes the slow changing of the spectrum to give the chloranil anion. An analogy can be drawn with the spectra of amino acid chloranil complexes which possess sharp bands in the region of 330 to 420 nm depending on pH and polarity of solvent, believed to be shifted bands of chloranil. On going over to the dimethyl-sulphoxide these bands are replaced by the chloranil anion bands (Section 3.3).

Retinoic acid chloranil spectra are a little different from those of retinol. On mixing in dimethyl formamide, a broad band appears at 570 nm which if a charge transfer band would suggest that retinoic acid has a higher ionization potential than retinol and is consequently a poorer charge donor. On dilution with buffer, a new sharp peak is observed at 316 nm. The addition of common salt does not cause the appearance of the chloranil negative ion. By analogy with the amino acids, retinol and retinoic acid would appear to be *n*-electron donors, retinol being the stronger donor.

The ability of related compounds to produce the TCNQ anion in methyl cyanide is β-carotene > retinol > retinoic acid > retinal (20).

9.4.2 IODINE

Colloidal retinol, but not retinoic acid, interacts with aqueous iodine to give a dark blue-green substance with an absorption maximum at 610 nm (21, 22). This substance, probably a retinol-iodine complex, decomposes with time and there is a concomitant increase in absorption at 226 nm from the iodine cation together with a new band at 870 nm. There are very obvious similarities with the interaction of the related compound β-carotene (Section 9.3.3) and the explanations put forward for the carotene iodine band in the near infra-red are applicable here (12, 13, 14).

In ethanol, both retinol and retinoic acid give the iodide spectrum but without the appearance of the dark blue-green complex as an intermediate. It is possible that the complex is only stable in water and not therefore observed in ethanol or that there exists two separate processes for transferring an electron to iodine to produce iodide, one operating in ethanol and the other in the presence of water.

9.4.3 OXYGEN

Molecular aggregates of retinol in anaerobic saline solution show a slight red shift in the absorption spectrum and oxidize very rapidly on the introduction of oxygen (23, 24). This is interpreted as arising from the production of retinol anions and cations due to charge transfer between retinol molecules in the presence of salt. The very rapid autooxidation of such aggregates occurs because the cation is far more reactive than the neutral molecule. There is a close parallel with the autooxidation of catecholamine which also procedes via the cation (Section 9.7.1).

9.5 Phenothiazines

This group of compounds is of great pharmacological interest as many of them are tranquillizers.

9.5.1 FLAVINS

The spectrum of a mixture of FAD and chloropromazine shows a complete loss of flavin absorption in the region of 450 to 500 nm (25) (see Fig. 9.13). The fluorescence of the flavin in the mixture is very strongly quenched. The association constant derived from quenching studies using the Stern-Volmer equation is *ca.* 1000 M^{-1}. Very similar results are found for complexes of RFN and FMN with the drug. The view that chlorpromazine forms a strong complex with FMN is supported by Karreman, Isenberg and Szent-Györgyi (26) who found at $-70°C$ a new absorption band at 490 nm which they believe to be the band of the semiquinone stabilized by charge transfer (Section 7.8.1) as well as a broad band at 570 nm similar to that exhibited by indole flavin complexes (Section 7.2).

9.5.2 PORPHYRINS

Cann (27) has demonstrated that chlorpromazine forms 1 : 1 complexes with hemin, haematoporphyrin and the protein myoglobin. 1 : 1 complexes are also formed with methylviologen.

9.5.3 ORGANIC ACCEPTORS

Foster and Hanson have examined the complexes formed by phenothiazines, including chlorpromazine, with a variety of organic acceptors (28). These mixtures exhibit charge transfer bands whose nature is established by the correlation of the position of maximum absorption of the bands with the different acceptors, and the positions of the bands observed for charge transfer complexes between the same acceptors and the strong donor hexamethylbenzene. The strength of complexing of these tranquillizers goes in the order chlorpromazine ≈ 10-methylphenothiazine < phenothiazine < 3,7-dimethylphenothiazine ≈ *N,N,N',N'*-tetramethylphenothiazinediamine.

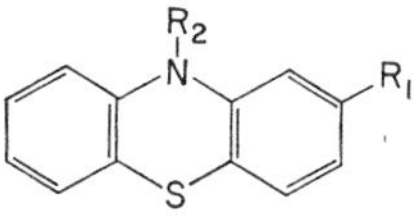

	R_1	R_2
1.	H	H
2.	H	$CH_2CH_2CH_2N(CH_3)_2$
3.	Cl	$CH_2CH_2CH_2N(CH_3)_2$
4.	H	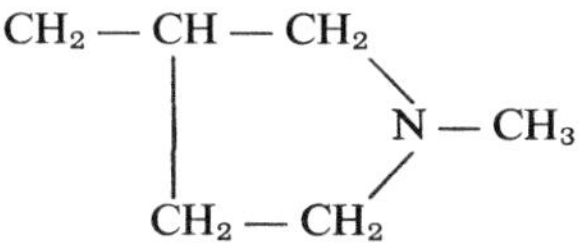 $CH_2 - CH - CH_2$ ... $N - CH_3$ ... $CH_2 - CH_2$
5.	H	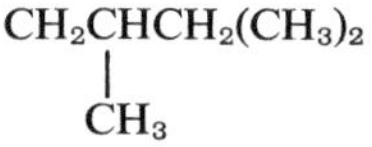 $CH_2CHCH_2(CH_3)_2$ with CH_3
6.	H	 $CH_2CHN(CH_3)_2$ with CH_3
7.	OCH_3	$CH_2CH_2CHN(CH_3)_2$
8.	Cl	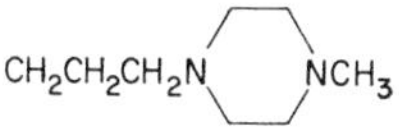 $CH_2CH_2CH_2N$⟨⟩NCH_3
9.	Cl	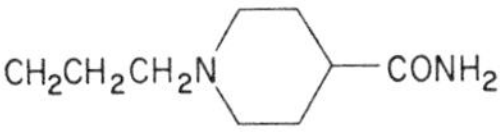 $CH_2CH_2CH_2N$⟨⟩$CONH_2$
10.	CF_3	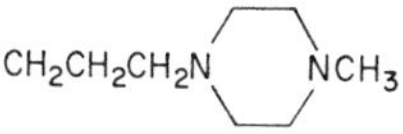 $CH_2CH_2CH_2N$⟨⟩NCH_3
11.	SCH_3	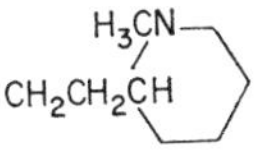 CH_2CH_2CH⟨⟩ with H_3CN-
12.	Cl	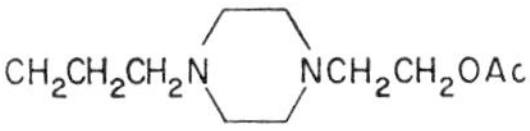 $CH_2CH_2CH_2N$⟨⟩NCH_2CH_2OAc

Fig. 9.10. (1) Phenothiazine, (2) promazine, (3) chlorpromazine, (4) methdilazine, (5) trimeprazine, (6) promethazine, (7) methoxypromazine, (8) prochlorperazine, (9) pipamazine, (10) stelazine, (11) thioridazine, (12) thiopropazate.

TABLE 9.7

Absorption maxima (kK) assigned as intermolecular charge-transfer transitions for mixtures of various heterocycles (A–F) with each of a series of electron acceptors in CH_3CN solution

| Acceptor | Donor[a] | | | | | |
	A	B	C	D	E	F[b]
(a) 2,3-Dichloro-5,6-dicyano-*p*-benzoquinone	—[c]	—	10.5	—	—	14.5	18.7
(b) 2,3-Dicyano-*p*-benzoquinone	—	—	11.9	—	—	16.5	21.3
(c) Tetracyanoethylene	11.7	10.9	12.1	—	11.8	16.4	20.9
(d) Bromanil	12.7	11.7	14.4	14.2	13.4	19.2	22.2
(e) Chloranil	12.7	11.8	14.4	14.4	13.7	19.4	22.7
(f) Trichloro-*p*-benzoquinone	12.9	11.8	14.3	14.1	13.4	19.4	22.7
(g) 2,6-Dichloro-*p*-benzoquinone	15.0	14.2	16.5	16.7	15.5	21.5	24.7
(h) 2,5-Dichloro-*p*-benzoquinone	15.2	14.9	16.3	16.6	15.7	22.2	—
(i) Monochloro-*p*-benzoquinone	16.9	16.1	18.3	18.8	16.9	23.5	26.0
(j) *p*-Benzoquinone	18.9	17.9	—	20.6	—	—	
(k) Methyl-*p*-benzoquinone	—	18.2	—	21.5	—	—	
(l) 1,3,5-Trinitrobenzene	20.2	18.6	22.7	—	19.5	—	
(m) *p*-Dinitrobenzene	21.0	19.6	23.2	—	20.6	—	
(n) 2,4,6-Trinitrotoluene	22.7	21.2	—	—	21.5	—	
(o) Iodine	—	—	—	—	—	26.7	

[a] Donors: A, phenothiazine; B, 3,7-dimethylphenothiazine; C, 10-methylphenothiazine; D, chlorpromazine; E, phenoxazine; F, thianthrene.

[b] Solvent for thianthrene complexes was dichloromethane.

[c] Dash indicates band not measurable, either due to chemical reaction or due to it being masked by absorption of the component molecules.

Reproduced with permission from Table 1, ref. 28.

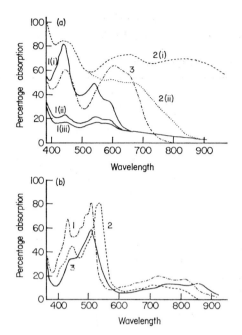

Fig. 9.11. (a) Absorption spectra in CH₃CN of: (1) 3,7-dimethylphenothiazine *plus* 2,3-dicyano-*p*-benzoquinone, showing the change in absorption with time (i → iii); (2) 10-methylphenothiazine *plus* 2,3-dicyano-*p*-benzoquinone, (i) in deoxygenated solvent showing the charge-transfer band, (ii) in normal solvent; (3) unsubstituted phenothiazine *plus* 2,3-dicyano-*p*-benzoquinone. (b) Absorption spectra in acidified CH₃CN of 2,3-dicyano-*p*-benzoquinone *plus*: (1) phenothiazine; (2) 3,7-dimethylphenothiazine; (3) 10-methyl-phenothiazine; showing the formation of free radicals. Reproduced with permission from Figs 1 and 2, ref. 28.

9.5.3.1 *Dinitrobenzene*

Association constants of a large number of charge transfer complexes existing between various phenothiazines and 1,4-dinitrobenzene in $CHCl_3$ and CCl_4 have been evaluated using the NMR method (29). The strength of complexing is of the order chlorpromazine < ethopromazine < promethazine < trimeprazine < diethazine < promazine < 10-methylpheno-thiazine.

9.5.3.2 *Chemical Interactions after Complexing*

A very comprehensive list of charge transfer band frequencies of many phenothiazine complexes with the acceptors tetracyanoquinodimethane, tetracyanoethylene, bromanil, chloranil, *p*-benzoquinone and *m*-dinitro-benzene have been presented by Fulton and Lyons (30). Even though these measurements were all taken in the non-polar solvent chloroform, the spectra

TABLE 9.8

Association constants (K_c) and chemical shifts (Δ_0) for protons in the acceptor moiety for complexes of various phenothiazine donor molecules with 1,4-dinitrobenzene in $CHCl_3$ solution and in CCl_4 solution at 33.5°

Donor	CHCl₃ solution			CCl₄ solution		
	K_c (kg/ mole)	Δ_0 (cycles/ sec)	Δ_{max}[a] (cycles- sec)	K_c (kg/ mole)	Δ_0 (cycles/ sec)	Δ_{max}[a] (cycles/ sec)
Chlorpromazine	0.205	143.0	25.5	1.19	58.6	28.1
Promethazine	0.238	153.0	30.6	1.32	72.2	35.2
Diethazine	0.288	131.7	32.9	1.28	63.0	41.6
Ethopropazine	0.230	167.1	33.5	1.25	76.0	41.1
Promazine	0.293	135.2	34.5	1.40	71.3	40.0
Trimeprazine	0.274	146.5	36.4	1.24	77.0	39.3
10-Methylphenothiazine	0.359	124.0	42.0	—	—	—[b]
Dimethylaniline	0.143	137.8	32.6	0.798	68.7	37.0

[a] Maximum shift observed experimentally.
[b] Too insoluble to be measured.
Reproduced with permission from Table 1, ref. 29.

of the acceptor anion were usually observed simultaneously with the charge transfer bands, thus illustrating the strong donor properties of many of these drugs. Many of these charge transfer complexes were intermediate states in chemical reactions. Those phenothiazines with secondary or tertiary amine groups appeared to form aminoquinones after complexing. The donated electron comes from the conjugated portion of the N-substituted phenothiazine. When the phenothiazine is in the form of the free base the terminal N of the substituent may act as an n-electron donor. Surprisingly the association constants of some of the complexes are very small, as are the apparent enthalpies of dissociation. The association constants are of the order of 1 M^{-1} and the enthalpies of dissociation range from -0.43 to -1.82 eV.

Ionization potentials of these drugs have been estimated from the positions of the charge transfer and maxima (30). These values are in conflict with the estimates of Slifkin and Allison using contact charge transfer spectra (31). Some of the charge transfer band maxima given by Fulton and Lyons (30) do not accord with unpublished data by the writer. For example Fulton and Lyons' value for the tofranil chloranil complex is 17.8 kK which the writer believes is a band from trichlorhydroxyquinone (32), the real charge transfer band having its maximum at 26.4 kK. Many of the charge transfer bands listed by Fulton and Lyons in the region of 18 kK for the chloranil complexes may well be due to the absorption of trichlorhydroxyquinone to which chloranil is very readily converted (32).

TABLE 9.9

Apparent equilibrium constants K and extinction coefficients ε of CT donor-acceptor systems in chloroform. Donor concentration range from 0.02 to 0.5 M. Values listed are means from six runs made at each temperature for each system

Donor-acceptor	Acceptor concn. (M)	K (l. mole^{-1})	ε	ΔH (eV)	$10^{-3}\nu_{CT}$ (cm^{-1})	Temp.
Tofranil-HCl -chloranil	2.551×10^{-3}	1.5 ± 0.2	$250 - 30$		16.3	20°
		1.0 ± 0.2	$250 + 30$	-0.99		30
		0.9 ± 0.1	$250 + 30$	$+0.05$		40
		0.9 ± 0.1	$250 + 30$			46
	4.439×10^{-3}	1.1 ± 0.1	240 ± 20		16.3	25
		1.0 ± 0.1	250 ± 20	-0.99		30
		0.9 ± 0.1	250 ± 20	± 0.05		40
		0.9 ± 0.1	240 ± 20			50
Tofranil-HCl -bromanil	6.592×10^{-4}	0.6 ± 0.2	1000 ± 330		16.0	20
		0.5 ± 0.1	670 ± 110	-0.65		30
		0.4 ± 0.1	670 ± 110	± 0.05		40
		0.4 ± 0.1	670 ± 110			50
	1.739×10^{-3}	0.4 ± 0.1	670 ± 110		16.0	25
		0.4 ± 0.1	670 ± 110	-0.65		30
		0.3 ± 0.1	670 ± 110	± 0.05		40
		0.3 ± 0.1	670 ± 110			50
Pertofran-HCl -chloranil	2.551×10^{-3}	0.8 ± 0.1	330 ± 50		16.5	20
		0.6 ± 0.1	330 ± 50	-1.34		30
		0.5 ± 0.1	330 ± 50	± 0.10		40
		0.5 ± 0.1	330 ± 50			46
	5.549×10^{-3}	0.7 ± 0.1	290 ± 30		16.5	25
		0.7 ± 0.1	290 ± 30	-1.34		30
		0.6 ± 0.1	230 ± 30	± 0.10		40
		0.5 ± 0.1	270 ± 40			50
Pertofran-HCl -bromanil	6.592×10^{-4}	0.8 ± 0.2	400 ± 80		15.9	20
		0.7 ± 0.2	400 ± 80	-0.48		30
		0.6 ± 0.1	400 ± 80	± 0.05		40
		0.6 ± 0.1	400 ± 80			50
	1.739×10^{-3}	0.7 ± 0.1	370 ± 40		15.9	25
		0.6 ± 0.1	370 ± 40	-0.43		30
		0.6 ± 0.1	370 ± 40	± 0.05		40
		0.6 ± 0.1	370 ± 40			50
Tegretol -chloranil	2.551×10^{-3}	1.6 ± 0.1	90 ± 10		20.8	20
		1.2 ± 0.1	110 ± 20	-1.82		30
		1.0 ± 0.1	130 ± 40	± 0.15		40
		0.9 ± 0.1	130 ± 40			46
Promazine-HCl -chloranil	5.610×10^{-3}	0.40 ± 0.07	630 ± 110		14.8	20
		0.38 ± 0.06	630 ± 110	-0.65		30
		0.36 ± 0.06	630 ± 110	± 0.05		40

Reproduced with permission from Table 3, ref. 30.

In dimethyl sulphoxide, the addition of some of these phenothiazines to chloranil gives the characteristic chloranil anion spectrum without any charge transfer bands (33). Similar results have been obtained by different authors for other groups of biomolecules (Sections 3.3 and 4.3).

9.5.3.3 Infra-red Spectrum of Chloranil Complex

The infra-red spectrum of the interaction product of phenothiazine with chloranil contains the spectrum of the chloranil negative ion (34). An important feature of this spectrum is that the carbonyl band of free chloranil

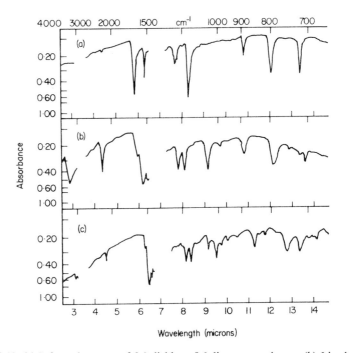

Fig. 9.12. (a) Infra-red spectra of 2,3-dichloro-5,6-dicyano-p-quinone; (b) Li salt of the emiquinone; (c) Infra-red spectrum of phenothiazine complex. Reproduced with permission from Figs 5 and 6, ref. 34.

is shifted from 1690 cm^{-1} to 1580 cm^{-1}. The presence of the anion spectrum is interpreted as being due to the presence of a charge transfer complex in which charge transfer is almost complete in the ground state rather than to the formation of free ions.

9.5.4 OXYGEN

9.5.4.1 *Polarography of Oxygen Complexes*

The half-wave potentials observed during the polarography of these drugs in aqueous solution show shifting on the addition of oxygen (35). These are interpreted as arising from the formation of complexes between oxygen and the drugs. No correlation is found between $\Delta E_{\frac{1}{2}}$, the half-wave potential and the wavelength of the first absorption band of the drugs. The first absorption band can be used as a rough measure of the ionization potentials in a conjugated π-electron system (36). It is concluded that these complexes do not involve the π-electron systems of these drugs. Additional confirmation comes from the order of shift in the $E_{\frac{1}{2}}$ values not correlating with the order of complexing found by Foster and Hanson (28). It is probable that the complexes with oxygen involve a σ electron on the carbon adjacent to the sulphur atom in the ring and is not therefore a charge transfer complex as defined in the first chapter.

9.5.4.2 *Contact Complexes with Oxygen*

In the presence of oxygen these drugs exhibit charge transfer spectra by which their ionization potentials have been measured (31). Bound complexes are not formed. These are examples of contact charge transfer.

9.5.5 XANTHENE DYES

Chlorpromazine xanthene dye mixtures have spectra uncharacteristic of the individual species (37). The new bands are designated as charge transfer bands. The fluorescence of the dyes is quenched by the addition of the drug. No ESR signal, denoting the presence of free radicals is observed, and it is concluded that chlorpromazine form charge transfer complexes with the dyes.

9.5.6 ELECTRICAL STUDIES

Several electrical studies have been carried out on thiazine complexes by Matsunaga (38). Dibenzophenothiazine can form 2 : 1 complexes with dicyanochloroquinone and 3 : 2 complexes with dibromocyanoquinone as well as the more usual 1 : 1 complexes. These complexes are dark green and are semiconductors with low resistivity. The infra-red spectra of these complexes suggests that they are essentially ionic in the ground state, *cf.* Section 9.5.3.3. The complexes with anomalous stoichiometry have a new absorption band at 800 nm not possessed by the 1 : 1 complexes. This is attributed to interaction between the ionized and neutral phenothiazine. These anomalous complexes have structures of the form $D_2^+ .. A^-$ and $(D_3)^{++} .. (A^-)_2$.

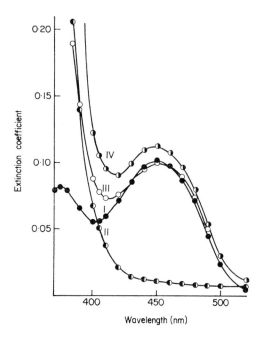

Fig. 9.13. Spectra of flavin adenine dinucleotide and chlorpromazine. I, Flavin adenine dinucleotide (9×10^{-6} M) in phosphate buffer (M/10, pH 6.5); II, chlorpromazine (2×10^{-3} M) in the same buffer; III, mixture of chlorpromazine (2×10^{-3} M) and flavin adenine dinucleotide (9×10^{-6} M) in the same buffer; IV, theoretical spectrum of flavin adenine dinucleotide + chlorpromazine. Reproduced with permission from Fig. 1, ref. 25.

Phenothiazine also forms dark-coloured complexes with iodine which are semiconductors with the unusual stoichiometry of 2 : 3 (39).

Similar studies by Guttman and Keyzer confirm that phenothiazine and chlorpromazine complexes iodine complexes are semiconductors with low resistivities (40). These complexes also give large ESR signals.

Conductimetric titration of solutions of phenothiazine or chlorpromazine with iodine in methyl cyanide indicate a 1 : 1 stoichiometry in freshly prepared solutions but a 2 : 3 stoichiometry with old solutions. Heated solutions show a 1 : 1 stoichiometry for the complex. The structure of the 2 : 3 complex is $I_2^- .. Ph^+ .. I_2 .. Ph^+ .. I_2^-$. The infra-red spectra of these complexes are identical to those published by Matsunaga (38). Gutmann and Keyzer point out that the spectra indicate that the interaction involves the whole of the phenothiazine nucleus rather than any specific sites. The spectra of these complexes closely resemble that of the phenothiazine alone at high temperature. This they suggest indicates the formation of a self-complex at these high temperatures. It is difficult to understand why high temperatures should favour self-complexing rather than low temperature as one would expect such a complex to have a negative enthalpy of dissociation.

Conductimetric titration of phenothiazines complexed with various quinones show them all to have 1 : 1 stoichiometry (41). There seem to be no indications of the unusual stoichiometries reported by Matsunaga (38), but these conductimetric studies were carried out in solution whereas Matsunaga measured stoichiometries in the solid state.

9.6 Menadione (Vitamin K_3)

Menadione is the quinone, 2-methyl-1,4-naphthoquinone. This compound being a quinone can be expected to behave as an electron acceptor.

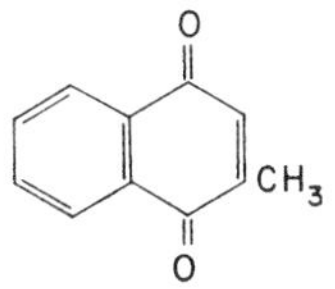

Fig. 9.14. Menadione (Vitamin K_3).

9.6.1 POLYCYCLIC AROMATIC HYDROCARBONS

Laskowski using the technique of fusing solids together and examining the solid-liquid phase diagrams, showed that some polycyclic aromatic hydrocarbons *viz*. hexamethylbenzene, pyrene and 3,4-benzpyrene form 1 : 2 molecular complexes with menadione and other quinones with vitamin K activity, which are orange or yellow (42). Other hydrocarbons, e.g. anthracene which didn't form complexes in the solid state nevertheless still gave yellow or orange colours in the molten mixtures.

Shatsung Liao and Williams-Ashton (43) have also shown that polycyclic aromatic hydrocarbons give red solutions when mixed with menadione. Furthermore these colours intensify on cooling. The stoichiometry of the complexes in solution is 1 : 1.

Association constants of polycyclic aromatic hydrocarbons menadione 1 : 1 complexes have been evaluated by several authors (44, 45, 46).

Cilento and Sanioto (46) have shown that solutions of mixtures of the polycyclic aromatic hydrocarbons with menadione in chloroform result in red-orange colours as shown by previous workers. Both association constants and thermodynamic parameters have been evaluated for the complexes. The association constants are very small, likewise the enthalpies of dissociation which in some cases are even positive. The authors explain these unusual positive results as probably arising from a large contribution of contact pairs to the spectrum. This fits in with the earlier work in which although colours

TABLE 9.10

Thermodynamic data for some complexes of aromatic hydrocarbons with menadione

Hydrocarbon	Solvent	T_1 °C	K_1 M^{-1}	T_2 °C	K_2 M^{-1}	ΔF_{T1} kcal/mole	ΔH kcal/mole	ΔS e.u.	$I_p^{\,a}$ eV
Anthracene	CHCl$_3$	26.2	0.55	49.2	0.62	0.35	1.0	2.1	7.37
Anthracene	CCl$_4$	24.6	0.12	49.2	0.95	1.26	large	large	
Tetracene	CHCl$_3$	10.0	5.27	51.4	14.2	−0.93	4.3	18.7	
Tetracene	CCl$_4$	19.4	2.2	48.8	1.5	−0.45	−2.3	−6.3	7.00
1,2,5,6-Dibenzanthracene	CHCl$_3$	26.0	0.50	48.6	0.63	0.41	1.9	4.9	
1,2,5,6-Dibenzanthracene	CCl$_4$	23.2	0.5–0.8	48.8	1.7–2.7	> 0	> 0	> 0	7.80

Adapted with permission from Table 2, ref. 46b.

[a] Birks, J. B. and Slifkin, M. A. (1961). *Nature*, **191**, 761.

occurred in the melt, these disappeared on solidification, which is explicable in terms of contact charge transfer which can only occur when donor and acceptor are specifically orientated with respect to each other (see Section 1.3).

TABLE 9.11

Data for complexes of polycyclic aromatic hydrocarbons with menadione in solution in $CHCl_3$ at 25°C ± 2°C

Hydrocarbon	λ nm	K_c (M^{-1})	I_p (eV)^a
7,12-dimethyl-1,2-benzanthracene	478	0.36	—
7-methyl-1,2-benzanthracene	476	0.40	7.37
10-methyl-1,2-benzanthracene	478	0.42	7.37
20-methylcholanthrene	500	0.43	
4-methyl-1,2-benzanthracene	474	0.45	7.41
1,2,5,6-dibenzanthracene	472	0.50	7.80
anthracene	472	0.55	7.37
2′-methyl-1,2-benzanthracene	476	0.60	7.39
3,4-benzpyrene	510	0.69	7.19
tetracene	484	5.27^b	7.00
3-methyl-1,2-benzanthracene	476	0.0	7.43
6-methyl-1,2-benzanthracene	472	0.4	7.37
3′-methyl-1,2-benzanthracene	472	0.4	7.43
1,2-benzanthracene	474	0.6	7.45

Correlation coefficient for K_c vs. I_p for unsubstituted hydrocarbons $= -0.87$ at 96% confidence level.

^{a} Birks, J. B. and Slifkin, M. A. (1961). *Nature*, **191**, 761.

^{b} At 10°C.

Adapted with permission from Tables 1 and 3, ref. 46b.

If in the solid state the molecules are unfavourably aligned then no charge transfer transition would occur.

There is some correlation between the association constants of the complexes and the ionization potentials of the hydrocarbons thus showing that there is some contribution to binding in the ground state from charge transfer.

9.6.2 TYROSINES

Full thermodynamic parameters have been worked out for the 1 : 1 complexes formed between menadione and tyrosine, 3,5-dibromotyrosine and 3,5-diiodotyrosine (47). Increasing halogenation of the tyrosine causes an increase in complexing ability. This is rather surprising in view of the electron withdrawing nature of the halogens. It is suggested by the authors that the halogen atoms might be directly involved in binding.

TABLE 9.12

Spectrophotometric and thermodynamic data for complexes of
menadione and tyrosine derivatives

Donor	λ nm	K_c (M^{-1})	ΔG (kcal/mole)	ΔH (kcal/mole)	ΔS (e.u.)	T (°C)	
tyrosine	380	2.93	− 0.57	− 2.47	− 6.34	26.0	a
MIT	380	26.64					b
DIB	400	12.10	− 1.48	− 2.64	− 3.86	27.1	a
DIB	500	5.12					b
DIT	380	19.84	− 1.77	− 3.27	− 7.51	26.0	a
DIT	500	7.56					b

MIT = 3-iodotyrosine; DIB = 3,5-dibromotyrosine; DIT = 3,5-diiodotyrosine.
[a] In aqueous HCl.
[b] In K_2CO_3–$KHCO_3$ solution.
Reproduced with permission from Tables 3 and 4, ref. 47.

9.6.3 DONORS OF PHARMACEUTICAL INTEREST

A series of experiments have been carried out on the complexes formed
between menadione and various biochemical electron donors of pharmaceutical interest by Hata and his colleagues (74–78).

TABLE 9.13

Properties of menadione complexes in water at 27°C

Donor	a	b	c	d	e	f	g
β-hydroxynaphthoic acid	4.49	6.52	− 7.23	2.34	77.2	0.384	2.79
salicyclic acid	3.48	4.23	− 2.76	1.61	71.7	0.438	3.26
Et aminobenzoate	3.28	4.00	− 2.66	1.20	60.3	0.363	—
theophylline	2.40	2.19	0.66	1.02	57.3	0.656	—
caffeine	2.22	2.00	0.57	1.00	53.4	0.633	3.18
dehydroacetic acid	2.07	0.92	3.71	0.96	47.7	0.822	3.32
nicotinamide	0.97	1.36	− 1.36	0.77	44.9	0.563	3.30
theobromine	—	—	—	0.62	43.6	0.633	—

Coefficients of correlation: a vs. f = − 0.66; a vs. e = 0.95; a vs. g = 0.78; b vs. e = 0.95;
b vs. f = −0.84; b vs. g = − 0.86; d vs. a = 0.92; d vs. b = 0.95; d vs. g = 0.87; f vs. g
= 0.64.

a = − ΔG (kcal/mol) ref. 74;
b = − ΔH (kcal/mol) ref. 74;
c = ΔS (entropy units) ref. 74;
d = calculated stability energy (molecular orbital coefficients) ref. 76;
e = photochemical stability constant (%) ref. 75;
f = energy of highest filled molecular orbital (molecular orbital coefficients) ref. 74;
g = energy of maximum charge transfer transition (eV) ref. 77.

The addition of the donors to menadione in water causes a decrease in absorption of the acceptor band but an increase in absorption to the long wavelength side (74). Association constants and thermodynamic parameters have been evaluated for the complexes as have the energies of the highest filled molecular orbitals of the donors. Consequently it is believed that these are charge transfer complexes.

Menadione decomposes on exposure to near ultra-violet light. The addition of the electron donors to aqueous menadione solutions helps stabilize the acceptor (75). The stabilization is proportional both to the free energy of the complexes and the enthalpy of dissociation. It is suggested that the stabilization of menadione against photodecomposition is due to the formation of charge transfer complexes.

Molecular orbital calculations on the complexes give energies of stabilization which linearly correlate with the experimental enthalpies of dissociation (76).

The charge transfer maxima of the complexes in aqueous solution linearly correlate with thermodynamic parameters and the stabilization against photodecomposition (77).

TABLE 9.14

Data for complexes of methylviologen

Donor	$K_c (M^{-1})$	λ_{max} (nm)	T (°C)	
N-phenyl-2-naphthylamine	88	550	20	a
ferrocyanide	52	~ 600	23	b
methylviologen ferrocyanide	0.21		23	b
hydroquinone $\Delta H = -5.45$ kcal/mole	5.23	440	25	c
pyrene	3.4	465	25	c
p-aminophenol	7.7	495	25	c
o-phenylenediamine	14.1	525	25	c
a-naphthylamine	8.7	525	25	c
p-phenylenediamine	15.4	555	25	c
chlorpromazine $\Delta H = -3$ kcal/mole	15	500	28	d
3-indoleacetate $\Delta H = -2.5$ kcal/mole	6	390	28	e
3-indolebutyrate $\Delta H = -2.5$ kcal/mole	13.5	400	28	e
a-naphthoate	9.1	375	28	e
a-naphthol	13.2	433	28	e

[a] Ref. 49, in methanol;
[b] Ref. 54, in water;
[c] Ref. 56, in methanol;
[d] Ref. 27, in aqueous buffer;
[e] Ref. 57, in pH6 buffer.

The effect of solvent of the charge transfer maxima has been investigated (78). The charge transfer transitions move to the red on increasing the polarity of the solvent, yet further evidence for the role of charge transfer forces in these complexes.

9.6.3 N-PHENYL-2-NAPHTHYLAMINE

Complexes of menadione and closely related quinones with a strong electron donor, the antioxidant N-phenyl-2-naphthylamine, have been demonstrated to exist by Ledwith and Iles (49) and the association constants measured spectrophotometrically. The menadione complex has $K_c = 4.5$ (M^{-1}) in dichloroethane.

9.7 Catechol

Catechol is a quinol containing two hydroxyl groups and it thus is an analogue of hydroquinone, a strong electron donor.

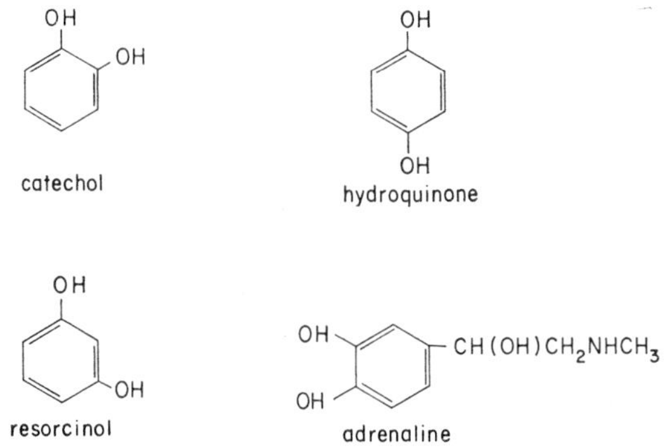

Fig. 9.15. Catechol and some related molecules.

9.7.1 OXYGEN

The spectrum of hydroquinone in solutions containing catechol changes fairly rapidly to the spectrum of its semiquinone (50). This change is greatly enhanced if oxygen is bubbled through the mixed solution. Bubbling oxygen through an ethanolic solution of catechol causes a new featureless absorption to appear similar to that seen for other biomolecules in the presence of oxygen (48) and is attributed to charge transfer between the catechol as donor and the acceptor oxygen.

It is thought that aerobic oxidation of the quinone is catalysed by the catechol which is activated by the formation of a charge transfer complex with oxygen. The donation of one electron to oxygen during complex formation enhances its electron affinity and thus makes it a better oxidizing agent in the complexed state than in the free molecular state. However, other work on the charge transfer spectra observed with oxygen point to no stable complexes being formed, only to there being momentary exchange of charge when the donor comes into contact with oxygen (48).

The dehydrogenation of the coenzyme model 1-benzyl-1,4-dihydronicotinomide pyridinium salt is catalysed by catechol in the presence of oxygen (50).

Other catechol-containing compounds which catalyse oxidation in the presence of molecular oxygen are the catecholamines, adenaline and noradrenaline (51). These amines convert the strong electron-donor, hydroquinone, to its cation in the presence of molecular oxygen. These are catalysed reactions as the catacholamine concentration appears to be unimportant to the speed and extent of the interaction. During the oxidation of hydroquinone, the spectrum of the mixed solution shows that the noradrenaline is oxidized to adenochrome so that the catecholamines are chemically changed during the reaction. This contradiction is resolved by a kinetic scheme in which one catecholamine catalyses its own oxidation and a second catecholamine catalyses the oxidation of the substrate thus:

$$\text{catecholamine}^- + O_2 \underset{k_2}{\overset{k_1}{\rightleftharpoons}} (\text{catecholamine}^- .. O_2)_{\text{complex}}$$

$$(\text{catecholamine}^- .. O_2) + \text{catecholamine}^-$$
$$\rightleftharpoons \text{catecholamine}^- + O_2^- + \text{catecholamine}.$$

the oxygen is now left in an activated state to act on the substrate.

One interesting aspect of these interactions are that they are specific. The oxygen catechol system has no effect on another closely related quinol of biological interest, resorcinol. The interaction of catechol with flavins does appear to be non-specific and is no different from the interactions of other phenols (Section 7.3).

The monoprotonated forms of p-phenylene diamines, good n-electron donors, activate oxygen in a similar manner to catechol (52).

9.7.2 NICOTINAMIDE AND ACETYCHOLINE

Complexes between catecholamines and acetycholine and nicotinamide have been investigated using a variety of techniques (53). Complexing causes oxidation of the catecholamines. Thus noradrenaline is oxidized to noradrenochrome, identified as such by the infra-red spectrum of the evaporated mixtures. Association constants for the noradrenalin acetylcholine and nicotinamide complexes have been obtained by monitoring the rate of formation of the complexes and are 5.7×10^3 M^{-1} and 7.1×10^2 M^{-1} respectively. Complex formation greatly increases the rate of autooxidation, nicotinamide having much the greater effect. These complexes also undergo chemical changes resulting in the formation of a fluorescent end product. The addition of ascorbic acid to the mixtures greatly increases the formation of this end product with acetylcholine but decreases the amount of end product with nicotinamide. Although these complexes have not been designated as charge transfer complexes, charge transfer must be occurring to produce the oxidized catecholamine. The results should be compared with those just discussed previously. The major difference in interpretation is that the complexes with acetylcholine and nicotinamide are stated to be with the oxidized catecholamine whereas in the scheme of Cilento and Zinner (52) the catecholamine complexes with oxygen and this then is catalysed by a further catecholamine to produce the chemical reactions. Such a scheme could be invoked to explain the present results.

9.8 Paraquat (Methylviologen)

Paraquat (methylviologen) has interesting biochemical properties. It inhibits electron transfer both in the cytochrome chain of mitochondria and

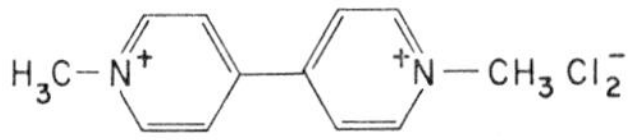

$$H_3C-\overset{+}{N}\!\!\!\diagdown\!\!\!\diagup\!\!\!\diagup\!\!\!\diagdown\!\!\!\overset{+}{N}-CH_3 \; Cl_2^-$$

Fig. 9.16. Paraquat.

in the electron transfer chain of chloroplasts. It is in widespread use as a weed killer.

9.8.1 SALTS

The absorption spectra of various paraquat salts show bands which are assigned to intermolecular transfer between the anion and the cation (54, 55). ESR studies on these salts lead to the conclusion that complete electron transfer occurs in some at least of these salts (55). Additional evidence for this is the high electrical conductivity of these salts.

9.8.2 ELECTRON DONORS

Complexes between paraquat dichloride and various π and n-electron donors have been investigated with both ultra-violet and NMR spectroscopy (31, 49, 55, 56). The paraquat salt forms 1 : 1 complexes with the donors. Charge transfer bands are observed and association constants have been evaluated using the Benesi-Hildebrand equation. The association constants of hydroquinone paraquat dichloride complexes were also evaluated by NMR methods and are identical to those obtained spectrophotometrically. The enthalpy of dissociation of the hydroquinone complex is 5.45 kcal/mole. In the solid state isolated complexes have 2 : 1 stoichiometry. This difference of stoichiometry between the solution and solid state is not too unusual. Similar differences have been commented upon in for example the phenothiazine complexes (Section 9.5.6). In solution the forming of a 1 : 1 complex lowers the electron positivity of the paraquat making it in the complexes state a poor electron acceptor. In the solid, the crystalline forces help to stabilize the complex in a 1 : 2 mode.

Positive identification of the charge transfer bands has been made by the very linear relationship between the position of the bands of the paraquat complexes and of the charge transfer complexes of the same donors with p-benzoquinone, chloranil, trinitrobenzene and tetracyanoethylene (56). The electron affinity of paraquat computed by correlating the charge transfer band positions with the energies of the calculated highest occupied molecular orbitals of the donors is 1.24 eV (see Table 9.14).

9.8.3 Complexing Strength from Spectral Shifts

Paraquat forms complexes with naphthoic acid and indole derivatives (57). In all cases new bands occur which can be identified as charge transfer bands as the mixtures obey the Benesi-Hildebrand equation and are reversible on dilution. The complexing strength judged by the association constants correlate very well with the shift of an intrinsic band of the donor in the complex. The association constants also show a very good correlation with the enhancement of urea denaturation of myoglobin by these molecules. The use of an intrinsic band of the donor to assess the amount of charge transfer complex whilst rather novel with biomolecules, is well accepted with organic complexes as for example iodine (Section 1.12) (see Table 9.14).

9.8.4 Chlorpromazine

The interaction of paraquat with chlorpromazine has been referred to earlier (Section 9.5.2).

9.8.5 N-Phenyl-2-Naphthylamine

Paraquat complexes with the antioxidant N-phenyl-2-naphthylamine (49).

9.9 Actinomycin

The antibiotic actinomycin complexes with purine derivatives accompanied by spectral changes in the visible (59, 60, 61). Kersten (59) has shown that there is an isosbestic point in actinomycin C deoxyguanosine mixtures. Using the spectral changes as a measure of complexing ability, it is concluding that the order of complexing of some purine derivatives with actinomycin C is guanosine > adenine > adenosine ≈ deoxyadenosine > AMP, ADP,

TABLE 9.15

Electron-donating properties of purines

Ring system	Energy coefficient K of the highest occupied molecular orbital
Thioguanine	0.16
Guanine	0.31
2,6-Diaminopurine	0.40
6-Dimethylaminopurine	0.41
6-Methylaminopurine	0.45
Aminopyrazolopyrimidine	0.49
Adenine	0.49
2-Fluoroadenine	0.49
2-Aminopurine	0.49
Benzimidazole	0.64
6-Methylpurine	0.66
Purine	0.69

Reproduced with permission from Table 1 of ref. 61.

ATP > inosine > xanthosine. This roughly correlates with the energy of the highest filled molecular orbital as calculated by Pullman (62). A very similar order of complex ability has been given by Pullman (62) from an examination of similar data of Reich (60), namely thioguanine > guanine > 2,6-diamino-

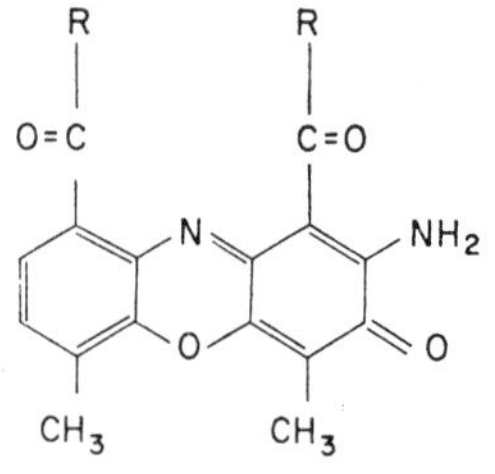

actinomycin

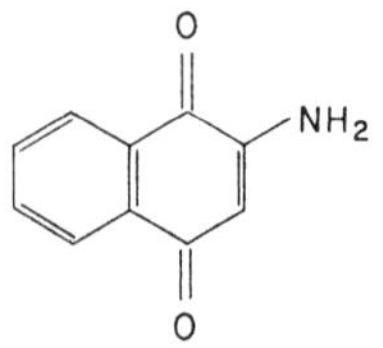

2-amino, 1,4-naphthoquinine

Fig. 9.17. Structures of actinoycin and 2-amino-1,4-naphthoquinone. R represents cyclic peptide sequences.

purine $\approx$ 6-dimethylaminopurine $\approx$ aminopyrazolopyrimidine > adenine $\approx$ 2-fluoradenine > 2-aminopurine $\approx$ benzimidazole > 6-methylpurine > purine. In view of the correlation of the complexing ability with the energy of the highest filled molecular orbital, Table 9.15, it is probable that stabilization is accomplished by charge transfer forces.

Behme and Cordes (63) have found that 1 : 1 complexes are formed between actinomycin D and derivatives of adenine and guanine, and have measured the association constants of these complexes. They express the opinion that stabilization is due in part to charge transfer forces. The association constants show the same order as those found by Kersten (59) and Pullman (62).

TABLE 9.16

Equilibrium constants for association of actinomycin D with derivatives of adenine and guanine. Measurements carried out spectrophotometrically at 25° and ionic strength 0.04 maintained with the addition of KCl. Potassium phosphate buffer (pH 6.9) and, for those measurements involving magnesium ion, imidazole buffer (pH 7.0) were employed for maintenance of constant pH. Except for those measurements involving magnesium ion, solutions contained 2.10^{-4} M EDTA. For those measurements involving magnesium ion, the concentration of metal ion was maintained at four times the concentration of total purine nucleotide. Actinomycin D concentration was approx. 4.10^{-5} M

Purine derivative	Concentration range (mM)	Association constant $(M^{-1} \times 10^{-3})$	
		425 mμ	440 mμ
Adenine	0.3 – 4.0	0.20	0.11
Adenosine	1.3 – 7.8	0.07	0.09
Deoxyadenosine	0.8 – 6.4	0.17	0.15
Adenosine 5'-phosphate	0.9 –10.4	0.10	0.14
Deoxyadenosine 5'-phosphate	0.9 – 9.9	0.22	0.26
Guanosine	0.35– 1.65	0.33	0.45
Deoxyguanosine	0.2 – 4.0	1.73	1.76
Guanosine 5'-phosphate	0.2 – 7.8	0.24	0.24
Deoxyguanosine 5'-phosphate	0.2 – 8.3	1.80	1.74
Guanosine 5'-phosphate plus Mg^{2+}	0.3 – 5.7	0.12	0.12
Deoxyguanosine 5'-phosphate plus Mg^{2+}	0.3 – 4.9	1.55	1.41

Reproduced with permission from Table 1, ref. 63.

Martin (64) has looked at the molecular extinction coefficients of bands occurring on adding 2-amino-1,4-naphthoquinone, believed to act as an analogue for part of the chromophoric centre of actinomycin, to various purine nucleosides but was unable to explain the results in terms of 1 : 1 interaction. Instead the results were interpreted as arising from non-specific hydrophobic interactions between a naphthoquinone and several purine molecules.

9.10 Procaine and Related Molecules

9.10.1 PURINES

The spectrum of mixtures of procaine hydrochloride and purines exhibit a strong sharp band at 340 nm with a temperature dependent intensity (65). It is suggested that this is indicative of a donor acceptor interaction with the C.CCO group as the acceptor. Such behaviour is quite unlike that normally associated with donor acceptor interactions.

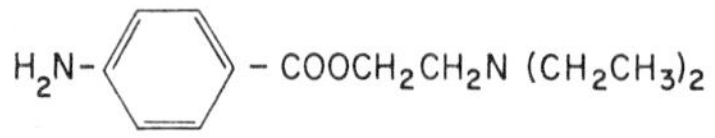

$$H_2N-\langle \bigcirc \rangle - COOCH_2CH_2N\ (CH_2CH_3)_2$$

Fig. 9.18. Procaine.

Agin has stated that the bands observed by Eckert (65) are artefacts of his method but RNA does give a charge transfer band with a maximum at 540 nm (66).

The association constants of complexes of ATP and procaine related local anaesthetics have been evaluated (67).

TABLE 9.17

Association constants of ATP with local anaesthetics

Compound	$K_c\,(M^{-1})$
Dibucaine HCl	17
procaine HCl	7.6
cocaine HCl	6.3
butanilicaine phosphate	4.9
tolycaine HCl	4.8
lidocaine HCl	3.1
mepivacaine	3.1

Reproduced with permission from ref. 67.

9.10.2 ORGANIC ACCEPTORS

Mixed solutions of procaine hydrochloride and chloranil are a blue-violet colour (65). It is suggested that the colouration is due to the formation of a charge transfer complex. Similar 1 : 1 complexes have been observed with trinitrobenzene (79).

Foster and Fyfe (68) have examined the complexes formed with the electron acceptors, trinitrobenzene and tetrafluorodicyanobenzene, using NMR techniques. The difference in association constants with different solvents is attributed to solvent competition for the *n*-electron in the aliphatic amine moiety of the donor. These complexes are believed to be charge transfer complexes with the donors acting simultaneously both as π-electron donors from the benzene ring and *n*-donors from the amine group. The donor is thought to bend round the acceptor so that each electron donating centre is on opposite sides of the acceptor.

9.11 Thiamine (Vitamin B$_1$)

The formation of 2 : 1 complexes between thiamine and acetylcholine, norepinephrine, serotonin and choline has been observed with differential ultra-violet spectrophotometry (69, 70). The principle effect of the additives to thiamine is a small blue-shift of the thiamine absorption band. No charge transfer bands are detected.

The conductivity of the mixtures is not the sum of the conductivities of the components, further evidence for charge transfer complexing. A molecular orbital study of the molecules reveals that there is good overlap between charges on atoms in the 2,3 position of the thiazole ring, the methylene group and the 4,5 atoms of the pyrimidine ring of thiamine with acetylcholine in its

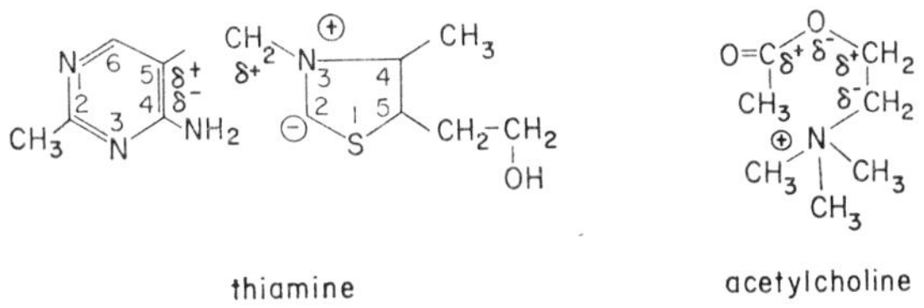

thiamine   acetylcholine

Fig. 9.19. Electronic structure of acetylcholine and thiamine. Reproduced with permission from ref. 69.

nicotinic form. Thus charge complimentarity can be evoked to explain the stability of the acetylcholine thiamine complex and probably the other thiamine complexes.

Eckert (71) has found that mixtures of thiamine and various local anaesthetics related to procaine form complexes which are characterized by new absorption bands (Table 9.18). The interaction with indoles is discussed in Section 5.12.

TABLE 9.18

Charge transfer bands of thiamine complexes with some local anaesthetics

Compound	Wavelength of maximum absorption
procaine	340
tetracaine hydrochloride	359
p-butylaminosalicylic acid 2-diethylaminoethyl ester	356
benoxinate	356
stadacain	301
cocaine	296
butanilicaine phosphate	281
dibucaine hydrochloride	359

Reproduced with permission from ref. 71.

TABLE 9.19

Association constants of local anaesthetics and related compounds
with organic acceptors

Donor	Solvent	Association constants (M^{-1})	
		Trinitrobenzene	Tetrafluoro 1,4-dicyanobenzene
benzocaine	$CHCl_3$	0.5_1	0.5_4
procaine	CCl_4	3.4	6.2_7
procaine	$CHCl_3$	0.5_6	0.6_4
lignocaine	CCl_4	9.6	11.1
lignocaine	$CHCl_3$	0.7_0	
prilocaine	CCl_4	8.0	7.5_3
prilocaine	$CHCl_3$	0.7_9	0.5_9
aniline	CCl_4	2.3_8	4.1_8
aniline	$CHCl_3$	0.5_4	0.5_6
2,6-dimethylaniline	CCl_4	3.5	4.3_5
aniline	$CHCl_3$	0.8_2	0.80

Reproduced with permission from Table 12.5, ref. 68.

9.12 Acetylcholine

During studies of neuromuscular action, Galzigna and coworkers have shown that acetylcholine forms 1 : 1 complexes with a variety of biochemicals, including the alkaloid, ajmaline (72), thiamine (69) see Section 9.11, chlorani-(69) and diazepam (70). In all cases the complexing was detected by perturbal tion shifts of the electronic absorption spectra. With the exception of chloranil no charge transfer bands were detected. The association constants where measured were in all cases very small indeed.

Further evidence is given for the enhancement of the conversion of catecholamine to noradenochrome in the presence of acetylcholine or nicotinamide by complex formation (73), *cf.* Section 9.7.2. Visible and infra-red spectroscopy of these systems suggest that the complexes are formed by the interaction of the $^-O-C-C-N^+$ and $^+O-C-C-N^-$ systems of the acetylcholine and adenochrome molecules, i.e. charge complimentarity. No charge transfer bands are observed however, *cf.* Section 9.11.

9.13 Aminopyrine

Mixed solutions of the antipyretic aminopyrine (4-dimethylamino-2,3-dimethyl-1-phenyl-3-pyrazolin-5-one) with benzoic acid or salicylic acid are coloured and have anomalous solubilities (80). Thermodynamic parameters and association constants have been evaluated (81). It is shown that aminopyrine forms 1 : 1 complexes, assumed by the authors to be charge transfer complexes (81). There is good agreement between the thermodynamic parameters found both by solubility and spectrophotometric methods. A difference in association constants, Table 9.20, is wrongly ascribed to the Benesi-Hildebrand method measuring charge transfer forces only whereas

TABLE 9.20

Complexes of acetylcholine

Compound	$K_c\,(\mathrm{M}^{-1})$	Solvent	Ref.
adenochrome	10^{-3}	chloroform	PS 18
adenochrome	1.75×10^{-4}	Tris-dioxan	[a]
chloranil	233×10^{-6}	unstated	PS 19

[a] Galzigna, L. (1970). *Nature*, **225**, 1058.

TABLE 9.21

Properties of aminopyrine complexes

Donor	ΔG	ΔH	ΔS	K_c	Temp	
Benzoic acid	-1.22	13.4	49.2	7.79	25	Solubility method
	0.55	12.9	43.4	0.91	25	Benesi-Hildebrand method
Salicylic acid	1.95	7.71	32.4	27.5	25	Solubility method
	-2.66	13	44.4	1.57	25	Benesi-Hildebrand method

Data from ref. 81.

the solubility methods measures all interaction forces. The difference between the results merely reflects the inadequacies of the methods used. The enthalpies of dissociation of the complexes are positive and hence any charge transfer forces must be negligible. A similar situation has been observed with the hydrogen-bonded complexes formed between aliphatic amino acids and chloranilic acid (82), which also has positive enthalpies of dissociation. This has been explained by the following scheme:

$$AOO^- + BOH \rightleftharpoons BO^- \ldots AOOH \rightleftharpoons BO^- + AOOH,$$

heating the mixed solutions causes the equilibrium to go over to the right thus giving the appearance of positive enthalpies of dissociation. It is quite possible that the complexing of the antipyrine to the weak acids is due to hydrogen bonding rather than charge transfer. The very high entropy changes on complexing suggests that perhaps solvation takes place, which might also be an explanation of the anomalous results.

9.14 Hallucinogens

Molecular orbital studies of various hallucinogens including those related to LSD (d-lysergic acid diethylamide) and those related to the alkaloids such as mescalin have been carried out by Snyder and Merril (83). A good correlation was noted between the hallucinogenic activity of the various drugs and the energy of the highest filled molecular orbital. It is suggested that the hallucinogens may operate by electron donation to some hypothetical acceptor.

REFERENCES

1. Allison, A. C., Peover, M. E. and Gough, T. A. (1962). *Life Sci.* 729.
2. Allison, A. C. and Nash, T. (1963). *Nature,* **197,** 758.
3. Szent-Györgyi, A. (1963). *Life Sci.* 112.
4. for example
    Bist, H. D. and Person, W. B. (1969). *J. phys. Chem.* **73,** 482; and references cited therein.
5. Wobschall, D. and Norton, D. A. (1967). *Archs Biochem. Biophys.* **122,** 85.
6. Bersohn, R. and Isenberg, I. (1961). *J. chem. Phys.* **35,** 1640.
7. Jones, J. B., Bersohn, M. and Neice, G. C. (1969). *Nature,* **211,** 309.
8. Bhowmik, B., Jendrasiak, G. L. and Rosenberg, B. (1967). *Nature,* **215,** 842.
9. Rosenberg, B. and Bhowmik, B. B. (1968). *Chem. Phys. Lipids,* **3,** 109.
10. Rosenberg, B. and Jendrasiak, G. L. (1968). *Chem. Phys. Lipids,* **2,** 47.
11. Slifkin, M. A. (1964). *Spectrochim. Acta,* **20,** 1391.
12. Lupinski, J. H. and Huggins, C. M. (1962). *J. phys. Chem.* **66,** 2221.
13. Lupinski, J. H. (1963). *J. phys. Chem.* **67,** 2725.
14. Huggins, C. M. and LeBlanc, O. H. (1960). *Nature,* **186,** 552.
15. Ebrey, T. G. (1967). *J. phys. Chem.* **71,** 1963.
16. Eley, D. D. and Snart, R. S. (1965). *Biochim. biophys. Acta,* **102,** 379.
17. Rosenberg, B., Misra, T. N. and Switzer, R. (1968). *Nature,* **217,** 423.

18. Lichti, F. U. and Lucy, J. A. (1969). *Biochem. J.*, **112**, 221.
19. Lichti, F. U. and Lucy, J. A. (1967). *Biochem. J.* **103**, 34p.
20. Foster, R. and Thomson, T. J. (1962). *Trans. Farad. Soc.* **58**, 860.
21. Lucy, J. A. and Lichti, F. U. (1967). *Biochem. J.* **103**, 34p.
22. Lucy, J. A. and Lichti, F. U. (1969). *Biochem. J.* **112**, 231.
23. Lucy, J. A. (1969). *Am. J. Clin. Nut.* **22**, 1033.
24. Lucy, J. A. (1965). *Proc. Biochem. Soc.* **96**, 12p.
25. Yagi, K., Ozawa, T. and Nagatsu, T. (1959). *Nature*, **184**, 892.
26. Karreman, G., Isenberg, I. and Szent-Györgyi, A. (1959). *Science*, **130**, 1191.
27. Cann, J. R. (1967). *Biochemistry*, **6**, 3427.
28. Foster, R. and Hanson, P. (1966). *Biochem. biophys. Acta*, **112**, 482.
29. Foster, R. and Fyfe, C. A. (1966). *Biochim. biophys. Acta*, **112**, 490.
30. Fulton, A. and Lyons, L. E. (1968). *Aust. J. Chem.* **21**, 873.
31. Slifkin, M. A. and Allison, A. C. (1967). *Nature*, **215**, 949.
32. Slifkin, M. A., Summer, R. A. and Heathcote, J. G. (1967). *Spectrochim. Acta*, **23A**, 1751.
33. Saucin, M., Van DeVorst, A. and Duschene, J. (1968). *Bull. Acad. Belg.*, **54**, 1006.
34. Matsunaga, Y. (1964). *J. chem. Phys.* **41**, 1609.
35. Martin, H. F., Price, S. and Gudzinowicz, B. J. (1963). *Archs Biochem. Biophys.* **103**, 196.
36. Birks, J. B. and Slifkin, M. A. (1961). *Nature*, **191**, 761.
37. Labos, E. (1966). *Nature*, **209**, 201
38. Matsunaga, Y. (1967). *J. chem. Phys.* **42**, 1982.
39. Matsunaga, Y. (1963). *Helv. Phys. Acta*, **36**, 800.
40. Guttmann, F. and Keyzer, H. (1967). *J. chem. Phys.* **46**, 1969.
41. Guttmann, F. and Keyzer, H. (1966). *Electrochim. Acta*, **11**, 555.
42. Laskowski, D. E. (1960). *Anal. Chem.* **32**, 1171.
43. Liao Shutsung and Williams-Ashman, H. C. (1961). *Biochem. Pharm.* **6**, 53.
44. Foster, R., Hammick, Ll. and Wardley, A. A. (1953). *J. chem. Soc.* 3817.
45. Fujimori, E. (1959). *Proc. natn. Acad. Sci.* **45**, 133.
46. Cilento, G. and Sanioto, D. L. (1963). (a) *Z. phys. Chem.* **223**, 333. (1963). (b) *Ber. phys. Chem.* **67**, 426.
47. Cilento, G. and Berenholc, M. (1965). *Biochim. biophys. Acta*, **94**, 271.
48. Slifkin, M. A. (1962). *Nature*, **193**, 464.
49. Ledwith, A. and Iles, D. H. (1968). *Chem. Brit*, **4**, 266.
50. Cilento, G. and Zinner, K. (1966). *Biochim. biophys. Acta*, **120**, 84.
51. Cilento, G. and Zinner, K. (1967). *Biochim. biophys. Acta*, **143**, 88.
52. Cilento, G. and Zinner, K. (1967). *Biochim. biophys. Acta*, **143**, 93.
53. Galzigna, L. (1970). *Nature*, **225**, 1058.
54. Nakahara, A. and Wang, J. H. (1963). *J. phys. Chem.* **67**, 496.
55. Macfarlane, A. J. and Williams, R. J. P. (1969). *J. chem. Soc.* (A), 1517.
56. White, B. G. (1969). *Trans. Farad. Soc.* **65**, 1.
57. Cann, J. R. (1969). *Biochemistry*, **8**, 4036.
58. Lovelock, J. E., Simmonds, P. G. and Vandenheuval, W. J. A. (1963). *Nature*, **197**, 249.
59. Kersten, W. (1961). *Biochim. biophys. Acta*, **47**, 610.
60. Reich, E. (1964). *Science*, **143**, 684.
61. Hamilton, L. D., Fuller, W. and Reich, E. (1963). *Nature*, **196**, 538.
62. Pullman, B. (1964). *Biochim. biophys. Acta*, **88**, 440.

63. Behme, M. T. A. and Cordes, E. H. (1965). *Biochim. biophys. Acta*, **108**, 312.
64. Martin, R. B. (1964). *Biochim. biophys. Acta*, **91**, 645.
65. Eckert, Th. (1962). *Arch. Pharm.* **295**, 233.
66. Agin, D. (1965). *Nature*, **205**, 805.
67. Eckert, Th. (1966). *Naturwiss.* **53**, 84.
68. Foster, R. and Fyfe, C. A. (1969). Unpublished work, quoted in Foster, R., "Organic Charge Transfer Complexes." Academic Press, London.
69. Galzigna, L. (1969). *Biochem. Pharmacol.* **18**, 2485.
70. Galzigna, L. (1970). *Int. J. Vitamin Res.* **40**, 38.
71. Eckert, Th. (1962). *Naturwiss,* **49**, 18.
72. Manani, G., Gasparetto, A., Bettinie, V., Caldesi Valeri, V. and Galzigna, L. (1970). *Agressologie*, **11**, 275.
73. Galzigna, L. and Rizzoli, A. A. (1969). *Clin. Chim. Acta*, to be published; Galzigna, L. *C.r. hebd. Séanc. Acad. Sci.* **286**, 2498.
74. Hata, S., Mizuno, K. and Tomioka, S. (1967). *Chem. Pharm. Bull.* **15**, 1791.
75. Hata, S., Mizuno, K. and Tomioka, S. (1967). *Chem. Pharm. Bull.* **15**, 1796.
76. Hata, S. (1968). *Chem. Pharm. Bull.* **16**, 1.
77. Hata, S. and Tomioka, S. (1968). *Chem. Pharm. Bull.* **16**, 1397.
78. Hata, S. and Tomioka, S. (1968). *Chem. Pharm. Bull.* **16**, 2078.
79. Strickland, W. A. and Robertson, L. (1965). *Proc. natn. Acad. Sci.* **54**, 452.
80. Regenbogen, A. (1918). *Chem. Zentr.* **2**, 625.
81. Okano, T., Uekama, K. and Ikeda, K. (1968). *Chem. Pharm. Bull.* **16**, 6.
82. Slifkin, M. A., Smith, B. M. and Walmsley, R. H. (1969). *Spectrochim. Acta*, **25A**, 1479.
83. Snyder, S. H. and Merril, C. R. (1965). *Proc. natn. Acad. Sci.* **54**, 258.

Postscript

Material in the postscript is to be read in conjunction with the text, as indicated by the section numbers.

3.7.1 EXCITED STATE STUDY OF LUMIFLAVIN AMINO ACID MIXTURES

Photolysis of aqueous solutions of flavins leads to the production of the semiquinone as a short-lived transient (1). The mechanism for this is believed to be the following (2); the absorption of light leads to the production of triplet excited flavin. In the case of lumiflavin, this can interact with ground state flavin leading to the semiquinone and another radical thus; $Fl^3 + Fl \rightarrow FlH^{\cdot} + Fl(-H)^{\cdot}$. Recombination of these radicals can take place after a brief interval of time.

The writer has very recently examined the effect of tryptophan and glycine on the photolysis of lumiflavin in water. The lifetime of the semiquinone of lumiflavin produced by photolysis is 3.2×10^{-4} sec (3). Neither the spectrum nor the lifetime is affected on initially adding the amino acid. In solutions more than a few days old however, although the spectrum is unchanged, the lifetime is greatly increased to $> 10^{-2}$ sec. This delay in interaction is presumably the same as explained in Section 3.4, in which the neutral amino acids form n-electron complexes with the ground state flavin which takes some time to reach equilibrium due to the slow conversion of the zwitterionic to the neutral form. The stabilization of the semiquinone can however be explained in several ways. The neutral amino acids could complex more strongly with the excited flavin which should be a much better acceptor than the ground state flavin. Whilst this might lead to a larger yield, it does not seem likely that it would lead to an increased lifetime. Alternatively the photolysis of the complex might lead to a dissociation of the complex with the semiquinone and an amino acid radical being produced. Again whilst this would give a larger yield, it is not certain that this would give an increased lifetime. A more probable explanation is that the amino acids complex with the $Fl(-H)^{\cdot}$ radical, which should be a good electron acceptor thus removing the quencher and hence increasing the semiquinone life.

3.11.1 EFFECT OF LIGHT ON D-AMINO ACID OXIDASE PURPLE COMPLEX

The purple complexes formed between D-amino acid oxidase and aliphatic amino acids or pipecolate are changed to a semiquinoid form of the flavin under illumination. The action spectrum of the reaction corresponds to the absorption spectrum of the purple complex (4).

4.3.1 COMPLEXES OF PURINES WITH p-BENZOQUINONE

Mixed solutions of p-benzoquinone and two purines, caffeine and theophylline, exhibit new absorption bands at 382 and 380 nm respectively (5). Spectrophotometric measurements indicate the formation of 1 : 1 complexes believed to be charge transfer complexes.

4.14.1 INTERACTIONS OF MIXED PURINES AND PYRIMIDINES

The association constants of complexes formed between mixed purines and pyrimidines in water at 25°C have been determined by Nakano and Igarashi (6) from phase-solubility diagrams. The authors ascribe the interaction to base stacking and do not suggest the presence of charge transfer forces. However, one can find some correlation and reverse correlation between the electron donor ability of the purines and pyrimidines. Thus methoxycaffeine has a higher association constant than caffeine in complexes with adenine or deoxyguanosine. On the other hand values for complexes for purine, thymine, cytosine and uracil are exactly in the reverse order to the association constants for complexes with RFN given Table 7.1. There are no associated changes in the absorption spectra of these complexes and therefore there is really no evidence to suggest the presence of charge transfer forces. It has nevertheless been shown in Section 4.14 that the luminescence spectra of some dinucleotides show changes that are reminiscent of charge transfer complexing. It may well be that luminescence studies of these mixed purines and pyrimidines would also show such changes (see Tables 4.7 and 4.12).

4.14.2 INTERACTION OF CYTIDINE WITH ITS CATION

The luminescence titration curves of frozen aqueous solutions of cytidine exhibit anomalies (7). Peak emission occurs at pH values very close to the pK value. This suggests that there is an interaction between cytidine and its protonated form as they are in equal concentration at pK = pH. At this pH the absorption spectrum stretches to longer wavelength than does the spectrum of either cytidine or its cation. Similarly, the fluorescence spectrum at this pH stretches to longer wavelength than do the individual spectra. These results are interpreted to mean that there is charge transfer interaction between cytidine and its cation both in the ground state and the first excited single state. In this complex the neutral cytidine is the electron donor and the cation the acceptor. The structure of cytidine is shown in Fig. 2 (p. 236).

5.5.1 INTERACTION OF INDOLES WITH p-BENZOQUINONE

Frozen solution of various indoles mixed with benzoquinone are coloured (11). The colour vanish on rewarming. The speed of this fading is affected by the presence of dissolved salts. It is presumed that charge transfer complexes of the quinhydrone type are formed.

5.16.1 HISTIDYL-TRYPTOPHAN PEPTIDES

Various histidyl-tryptophan peptides exhibit intense hyperchromaticity in the region of 240 nm which is attributed to an intermolecular imidazolium-indole complex (12). Fluorescence titration shows a marked effect with protonation of the imidazolium side chain. It is believed that charge transfer occur between the C2–C3 bonds of the indole (see Fig. 5.1) and the positively charged centre of the imidazolium. The fact that the electron acceptor is positively charged is believed to explain the very high value of the charge transfer maximum wavelength, as there will be no coulombic interaction term contributing to the energy of the charge transfer transition hv_{CT}.

Association constants of several intermolecular complexes between indole and various imidazolium and pyridinium compounds have been obtained from fluorescence quenching measurements. These association constants obtained from the Stern-Volmer equation are in general very much higher than those obtained from the Benesi-Hildebrand equation, $cf.$ Tables 5.8, 5.12 and 8.11. This indicates the presence of stronger complexing in the excited state than in the ground state (see Table 5.15).

5.17 Interaction between Indoles and Coenzyme Models

The 1 : 1 charge transfer complexes formed between various indoles and the coenzyme models, N-methylnicotinamide chloride and N-methyliso-nicotinamide chloride has been extensively studied by Deranleau and Schwyzer (13). The charge transfer band observed in the mixed solutions is believed to be the sum of two overlapping roughly symmetrical absorption bands. The presence of two bands is indicated by quantum-mechanical calculations which suggests that two electronic transitions are possible in these types of complexes (see Table 8.11).

The association constants, obtained by observing the absorption spectra of the mixtures on titrating the donor with the acceptor show a dependence on solvent composition ($cf.$ Section 7.18.4).

5.18 Interaction of Indoles with Flavins

A very thorough study has been made of the quenching of FMN fluorescence by various indoles at different temperatures (14). From the variation of the Stern-Volmer quenching coefficients at different temperatures, apparent activation energies have been obtained. These energies are not only of the same order as typical charge transfer complexes but increase on the addition of nucleophilic groups to the indoles. These results are consistent with the stabilization of the indole-flavin complexes by charge transfer forces.

The effect of water, or perhaps polarity, on these complexes has been examined and like other workers (Section 7.8.14) it is found that interaction between the components increases with increasing water content, or perhaps polarity of solvent.

The apparent association constant of the FMN-tryptophan complex has been measured from absorption spectrophotometry and found to be −6.6 kcal (*cf.* value of −7.4 in Table 7.2). The difference between the association constants obtained from absorption studies and those obtained from fluorescence quenching suggests the existence of both ground state and excited state complexes (see Tables 5.12 to 5.14).

5.20 Indole Exciplexes

Several authors have shown that the fluorescence of indole and its derivatives is red-shifted in the presence of polar solvent molecules (8, 9, 10). This is due to the formation of exciplexes, i.e. complexes between the excited indole and the polar solvent molecule. Although it is suggested that as well as 1 : 1 complexes 1 : 2 complexes are formed with alcohols (8, 9), Longworth (10) has been unable to repeat the results for those compounds thought to form 1 : 2 exciplexes if the systems are carefully dried. It is postulated that exciplexes are bound by charge transfer between the ground state polar molecule as donor and the excited indole as acceptor so that the wavefunction of the exciplex can be expressed as

$$\psi_E = a\psi(DA^+) - b\psi(D^+A^-) \qquad \text{with } b \gg a$$

in an analogous manner to the Mulliken formulation of charge transfer complexes (see Table 5.16).

6.10.2.1 *ESR of Porphyrin Quinone Systems*

Broad signals have been observed in irradiated porphyrin-benzoquinone mixtures in aerated media which are attributed to the formation of porphyrin-oxygen complexes (15, 16).

Light irradiated porphyrin quinone mixtures in polar solvents have been studied (17). The characteristic ESR signal of the semiquinone has been observed in N-methylformamide and N-ethylacetamide but not N,N-dimethylformamide nor N,N-diethylacetamide even when aerated. This suggests that the presence of the $N-H$ group lessens the influence of oxygen. The normal semiquinone signal is observed in deaerated N,N-dimethylformamide and N,N-diethylacetamide.

7.2.1 Intramolecular Interactions in Flavinyl Peptides

The intramolecular interactions in flavinyl peptides between the flavinyl and tryptophyl moieties has been studied using proton magnetic resonance spectroscopy (18). The results show that the interaction is mainly between the aromatic portion of the flavin and the amino acid. This it is believed shows that charge transfer complexing is not an important force in stabilizing the intramolecular complex in the ground state because calculations by Karreman (19) suggest that if there are predominant charge transfer forces then the benzenoid, and heteroaromatic portions of the tryptophyl should lie above the pyrimidine and heteroaromatic portions of the flavin.

7.4.1 Excited State Study of Lumiflavin Nucleotide Mixtures

The effect of some purine nucleotides on the excited state spectrum of lumiflavin is rather different to the effect of the amino acids, see Section 7.8.1.1 (20). One important difference in condition to be noted is that the studies were carried out in buffered solutions. The effect of the nucleotides is to decrease the yield of lumiflavin semiquinone and to protect the lumiflavin against photobleaching. The authors reject the idea of charge transfer complexing being important in these interactions as at the concentrations used, no change in either fluorescence or absorption spectra are seen. However, it is conceded that complexing with the triplet state is a possibility, although the authors believe that a quite different mechanism operates. These results are quite different to those of the immediate previous section but as shown by Vaish and Tollin (2), the reactions of lumiflavin in buffers is quite different from that in pure water.

7.4.2 Role of van der Waal's and Dispersion Forces
in Purine Flavin Complexes

Mantione (37) has calculated the interaction energies for complexes of purines with riboflavin. She has considered electrostatic polarization, dispersion and exchange interactions. Although she has stated that these account for the majority of the binding energy of these complexes, in fact the correlation coefficient assuming a linear regression between the total energies and the association constant of Tsibris et al. (Ref. 29 of Chapter 7 and Table 7.1) is only 0.81, whereas the correlation coefficient with the molecular orbital coefficients giving the ionization potentials of the purines is 0.87 (see Table 7.1). In addition, the interaction energies of dispersion and polarization forces include contributions from ionized forms of the molecules so that an even lower correlation coefficient would be obtained if these were excluded from Mantione's calculations.

7.5.1 EXCITED STATE OF FLAVIN AROMATIC HYDROCARBON COMPLEXES

Recently the writer has prepared solid complexes of riboflavin and the carcinogen 3,4,8,9-dibenzpyrene. The excited state spectrum has been searched for in specimens mulled in liquid paraffin. No signals of any kind were detected, although the hydrocarbon on its own gives a particularly strong signal (3). In solution also the effect of adding riboflavin to the hydrocarbon is to completely quench the triplet-triplet spectrum of the hydrocarbon. This confirms that there is an interaction between flavins and hydrocarbons although no light is thrown on the mechanism. One feature of charge transfer complexing is that the triplet-triplet transitions of the donor are very strongly quenched.

7.5.2 INTERACTION OF FLAVINS WITH VARIOUS AROMATIC MOLECULES

The association constants of the complexes formed between flavins in all three redox forms with various aromatic compounds including indoles and aromatic amino acids, have been determined with a potentiometric titration technique (21). The indoles complex quite strongly with all three forms of flavin. Phenylalanine does not appear to complex at all with the semiquinoid flavin as not do pyridine, nicotinic acid and picolinic acid (see Table 7.14 and 7.15).

7.5.3 MODIFICATION OF THE PHOTOCONDUCTIVITY OF ORGANIC MOLECULES

The addition of riboflavin to the aqueous electrode causes a marked increase in the photocurrent of crystalline anthracene (22). Electron transfer takes place between the ground state anthracene as donor and the light excited riboflavin as acceptor. No effect is observed with ground state riboflavin. It is expected that an excited acceptor will function as a better acceptor in the excited state than the ground state. This work adds further confirmation to the idea that the complexes formed between aromatic hydrocarbons and flavins (Section 7.5) could well be charge transfer complexes.

7.7.1 QUENCHING OF FLAVIN SEMIQUINONE BY DIMERIZATION

The quenching of flavin semiquinones produced either by electron pulse radiolysis or flash photolysis is thought to procede by dimerization thus:

$$FlH^{\cdot +}FlH^{\cdot} \rightarrow (Fl : FlH_2) \rightleftharpoons Fl - FlH_2$$

in other than pure water (23, 24).

7.8.1 FLAVOPROTEIN NAD COMPLEX

In a recent study of the role of NAD^+ in the catalytic mechanism of the flavoprotein lipoamide dehydrogenase (25), the spectra of the complexes between NAD^+ and 3-APNAD^+ with the flavoprotein have been given. In the latter case a band at *ca.* 700 nm called the charge transfer band is noted. In both cases bands at 507 and 387 nm occur in the difference spectra, and negative bands at 477, 450, 430 and 370 nm also occur. These spectra bear a superficial resemblance to those observed with flavin-electron donor complexes in the region 350 to 520 nm. The association constant for the NAD^+ complex is *ca.* 10^4 M^{-1} at 5°C.

7.8.2 INTERACTION OF CARBOXYLATES WITH D-AMINO ACID OXIDASE

Association constant and thermodynamic parameters for the purple complexes between D-amino acid oxidase and various carboxylates have been determined by Yagi *et al.* (26), Table 7.16 and 7.17. The dominant term in the enthalpy of dissociation is a positive entropy change which suggests that solvation effects, i.e. a decreased ordering of water molecules when the complex is formed, occurs. Complex formation is greatly facilitated by the presence of a nitrogen atom to the carboxylate group. It is suggested that complexing occurs in part from *n*-electron donation from the nitrogen in the amino group of the carboxylate to the isoalloxazine moiety of the flavoprotein.

7.9 Flavoproteins

There has been a very recent review of the chemistry and molecular biology of flavins and flavoproteins (27). Among the points made by the authors are the following. It is pointed out that whereas the free flavins can in principle form sandwich-like π-π complexes, this is not true of the flavoproteins which would need to undergo large conformational changes before this type of complex could form. On the other hand, other kinds of charge transfer interaction not involving π-π overlap are possible. The review stresses the role of the semiquinone in the molecular biology of the flavins.

7.9.11.1 *Lack of Intramolecular Complexing in FAD in Ethanol*

The phosphorescence lifetime of FAD in frozen ethanolic solution is very similar to that of RFN (28). This again shows the importance of water in the promotion of complexing of the flavins.

7.10 Inra-red Spectra of Caffeine and Tryptophan Flavin Complexes

Complexes of RFN with tryptophan and caffeine have been prepared by the writer by evaporating down equimolar aqueous solutions of the compounds. The resulting residues are very strongly coloured as in Fig. 7.2. The infra-red spectra of these residues have been determined in KBr discs. The only notable feature of these spectra is the shift of the RFN carbonyl band from 1740 to 1720 cm^{-1}. The band is due to the carbonyl in the 4 position (29). The carbonyl at the 2 position is unaffected. No changes in the infra-red spectra of either tryptophan or caffeine are noted. These results are consistent with the formation of weak charge transfer complexes. The spectrum of RFN alone in the charge donating solvent dimethyl sulphoxide shows the same changes. The result is inconsistent with that in Section 3.3 in which tryptophan is shown to behave as an n-electron donor, but the conditions are not strictly comparable. The fact that only one carbonyl groups is affected suggests that charge donation is occurring to a localized site. The Pullmans have shown theoretically that the 4 position in isoalloxazine is the most electron deficient region of the molecule (30).

7.11 Temperature Effects

Douzou (31) has presented various spectra of systems of FMN with tryptophan, methoxyindole and p-phenylene diamine as a function of temperature. The resulting changes are said to be due to variations in dielectric constant of the solvent and decreases in Brownian motion and viscosity. It is believed that very low temperatures aggregations of the form $(D^{\cdots\cdots}A)_n$ ($n = 2, 3$, etc.) occur. The stabilizing forces for these aggregations are thought to be van der Waals forces and dispersion forces as well as charge transfer forces.

8.2.2.1 *Flash Photolysis of the Charge Transfer Bands of Coenzyme Model Salts*

The earlier work of Kosower and Lindqvist (Section 8.2.2) has been confirmed by the flash photolysis of the charge transfer bands of various coenzyme model salts (32). The absorption of light by the salt results in the transfer of an electron from the cation to the anion thus:

$$R^+I^- \rightleftharpoons R^{\cdot}I^{\cdot}$$

which is followed by dissociation and reaction with the iodide cation to produce the iodine negative ion:

$$I^{\cdot} + I^- \rightleftharpoons I_2^-.$$

8.2.5.1 *Interaction of Reduced and Oxidized Coenzyme Models*

Coloured solutions are formed when 1-n-propyl-1,4-dihydronicotinamide and 1-n-propyl nicotinamide are dissolved together (33). This is due to the formation of a 1 : 1 charge transfer complex with an association constant of 0.89 M^{-1} and an extinction coefficient of 89 at 500 nm in pH 8.5 buffer at 24°C. No interaction was observed, in common with other workers between NAD$^+$ and NADH.

8.2.6.1 *Modification of the Photoconductivity of Aromatic Hydrocarbons*

The addition of various coenzyme model compounds causes a marked increase in the D.C. photoconductivity of organic molecular crystals (22). The effect of these compounds correlates very well with their electron-acceptation properties. The effect of these compounds is to apparently remove an electron from the excited organic molecule converting it into a better conductor by "hole" conduction. This type of process has obvious affinities with charge transfer complexing. It has been demonstrated (Section 8.2.6) that these hydrocarbons can form charge transfer complexes with similar coenzyme model compounds.

8.3.4 INTERACTION OF NICOTINAMIDE METHOCHLORIDE AND IODINATED PHENOLS

Solutions of various iodinated phenolic amino acids and nicotinamide or other coenzyme model salts, possess charge transfer bands (34). Association constants for these various complexes have been evaluated spectrophotometrically. The association constants are of very similar magnitude to the association constants of the indole complexes, cf. Tables 8.11, 8.12 and 8.13, although the wavelengths of maximum charge transfer absorption are very different. This does suggest that the interaction mechanism is the same both for the iodinated phenols and the indoles; probably π-electron transfer from the indole or phenol ring.

8.4.1 TEMPERATURE EFFECTS

Douzou (3) has investigated the effects of temperature on solutions of NAD with serotonin and 6 methoxy-indole spectrophotometrically. Decreases in temperature cause marked increases in optical density which are only partially thermally reversible. These are ascribed to aggregation of the components followed by oxidation. The aggregating forces are thought to be van der Waals forces, dispersion forces and charge transfer forces.

9.2.1 COMPLEXES OF EGG LECITHIN

The equilibrium constants of the 1 : 1 charge transfer complexes formed between egg lecithin and the organic acceptors have been determined spectrophotometrically (35). The association constants are in the order picric acid > m-chlorophenylhydrazone of carbonyl cyanide > trinitrobenzene. The stability of the complexes are related to the increase in electrical conductivity of bimolecular lipid membranes on complexing with the same acceptors (cf. Section 9.2).

9.5.3.4 Interactions of Tranquilizers with Organic Acceptors

Saucin and van de Vorst have looked at charge transfer complexes formed between various molecules of pharmacological interest, mainly of the phenothiazine type with different organic acceptors (36). In polar solvents the negative ions of the acceptors, i.e. chloranil, trinitrobenzene and tetracyanoquinodimethane are seen, whereas in non-polar solvents charge transfer bands occur (cf. Sections 3.3 and 4.3).

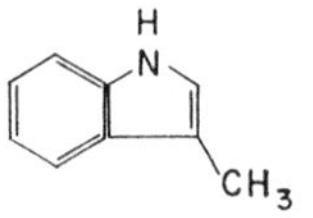

Fig. 1 (postscript). Skatole (β-methylindole).

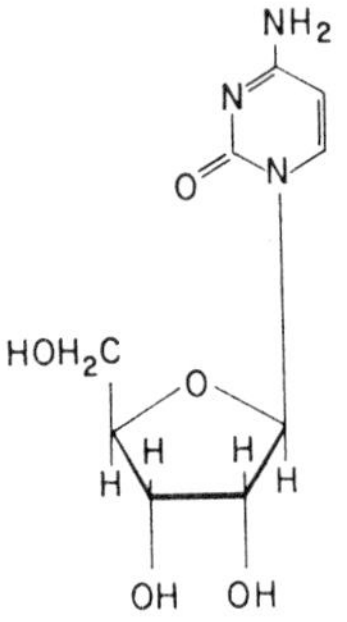

Fig. 2 (postscript). Cytidine. The cytidine cation is similar to cytidine but with a proton at the N_3 position.

TABLE 4.7

Adenine complexes[a]

Complexing agent	$K_{app\,1:1}$ M^{-1} at 25°
Uracil	4.16
5-Fluorouracil	5.44
5-Chlorouracil	6.33
5-Bromouracil	7.04
Uridine	4.03
5-Iodouridine	9.63
1,3-Dimethyluracil	7.14
Thymine	7.00
Cytosine	4.89
Caffeine	45.1
8-Methoxycaffeine	80.2
Theophylline	31.1
Purine	11.3
Adenosine	21.0
Deoxyguanosine	18.8
Inosine	8.25
Quinoxaline	17.8
Chloroquine	39.2

[a] Average solubility of adenine in water at 25° = (7.80 ± 0.078) × 10^{-3} M. Tables 4.7—4.12 are reproduced with permission, from ref. 6.

TABLE 4.8

Deoxyguanosine complexes[a]

Complexing agent	$K_{\text{app 1:1}}$ M^{-1} at 25°
Cytosine	3.32
1,3-Dimethyluracil	4.73
Caffeine	26.5
8-Methoxycaffeine	30.7
Theophylline	14.0
Purine	7.61
Adenosine	9.21
Inosine	6.92
Quinoxaline	14.3
Chloroquine	38.0

[a] Average solubility of deoxyguanosine in water at 25° = $(12.90 \pm 0.35) \times 10^{-3}$ M.

TABLE 4.9

Hypoxanthine, adenosine, and guanosine complexes

Complexing agent	$K_{\text{app 1:1}}$ M^{-1} at 25°
Hypoxanthine–caffeine	10.1
Adenosine–caffeine	39.4
Guanosine–caffeine	15.6
Guanosine–8-methoxycaffeine	23.0
Guanosine–quinoxaline	13.8

TABLE 4.10

Apparent 1:1 stability constants (M^{-1}) of alkylxanthine complexes at 25°

	Caffeine	8-Methoxycaffeine	Theophylline
Adenine	45.1	80.2	31.2
Adenosine	39.4		
Guanosine	15.6	23.0	
Deoxyguanosine	26.5	30.7	14.0
Hypoxanthine	10.1		

TABLE 4.11

Effect of 2′-hydroxl group on the interactive tendency of guanine nucleosides, $K_{app\,1:1}$ (M^{-1}) at 25°

	Guanosine	Deoxyguanosine
Caffeine	15.6	26.5
8-Methoxycaffeine	23.0	30.7
Quinoxaline	13.8	14.3

TABLE 4.12

Interactions of 5-halogenated uracils with adenine

Complexing agent	$K_{app\,1:1}$ M^{-1} at 25°
Uracil	4.16
5-Fluorouracil	5.44
5-Chlorouracil	6.33
5-Bromouracil	7.04
Uridine	4.03
5-Iodouridine	9.63

TABLE 5.12

Stern–Volmer constants for the quenching of FMN fluorescence by various indoles. Aqueous solutions in 0.05 M phosphate buffer (pH 7.0)

Compound	Temperature (°C)					Apparent activation energy (kcal/mole^{-1})
	10	25	40	55	70	
Indole	102	93	82	77	68	− 1.28
3-Methyl indole	153	132	102	78	75	− 2.47
5-Methyl indole	180	134	123	113	103	− 1.65
5-MeO indole	250	198	140	136	105	− 2.65
Indole-3-acetic acid	89	73	57	54	48	− 1.98
Indole-3-propionic acid	126	86	76	65	62	− 2.16
Indole-3-butyric acid	126	81	—	—	—	—
Gramine	172	136	102	83	62	− 3.68
Tryptamine HCl	219	146	143	95	64	− 3.61
N-dimethyltryptamine	212	166	133	109	98	− 2.69
N-Dimethylhomotryptamine	218	150	130	103	80	− 3.22
5-OH Tryptamine (oxalate)	500	400	260	200	158	− 3.78
Bufotenine (bioxalate)	450	300	225	185	148	− 3.59
5-MeO-N-dimethyltryptamine	570	370	255	177	146	− 5.01
5-MeO-tryptamine	640	400	280	214	185	− 4.00
N-acetyl-5-OH-tryptamine	375	270	215	158	142	− 3.11
L-Tryptophan	144	110	85	72	67	− 2.41
DL-Tryptophan	144	118	101	86	73	− 2.12
5-OH-tryptophan	352	230	178	161	129	− 3.02
N-Acetyl tryptophan	115	76	56	49	41	− 2.79

Reproduced with permission from Table 1, ref. 14.

TABLE 5.13

The effect of solvents on the quenching of FMN fluorescence by indole (25°C)

Solvent system	Stern–Volmer constant (l. mole^{-1})
Water	93
Ethanol–water (97.5 : 2.5 v : v)	35.7
Ethanol–water (50 : 50 v : v)	31.6
Methanol–water (97.5 : 2.5 v : v)	57.8
Methanol–water (50 : 50 v : v)	38.8
Glycerol–water (50 : 50 v : v)	30.4

Reproduced with permission from Table 2, ref. 14.

TABLE 5.14

The effect of temperature on the quenching of FMN by indole in different solvent

Solvent system	Quenching constant (l. mole^{-1})					Apparent activation energy (kcal. mole^{-1})
	10°C	25°C	40°C	55°C	70°C	
Ethanol–water (97.5 : 2.5 v : v)	29.9	35.7	40.2	46.2	48.5	1.75
Methanol–water (97.5 : 2.5 v :)	47.9	57.8	60.9	65.6	74.2	1.33

Reproduced with permission from Table 3, ref. 14.

TABLE 5.15

The Stern–Volmer constant K_{SV}, of the quenching of the fluorescence of 1 μM indole by various imidazolium and pyridinium derivatives (1–10 mM) in water. Limits of errors are estimated

Quencher	KSV
	l/mole
Imidazole · HCl	13 ± 1
Histamine · 2 HCl	17 ± 1
a-N-Acetyl-L-histidine amide · HCl	28 ± 1
Pyridine · HCl	75 ± 1
1-Methyl pyridinium chloride	90 ± 2
1-Methyl-3-carboxamide pyridinium chloride	142 ± 3

Reproduced with permission from Table 1, ref. 12.

TABLE 5.16
Association constant of indole exciplexes

Indole	Solvent molecule	K (M^{-1})
indole	diethyl ether	0.20
	dioxan	0.28
	acetonitrile	1.21
	butanol	109.2[a] (M^{-2})
	methanol	141.3[a] (M^{-2})
1-methylindole		
	diethyl ether	0.14
	dioxan	0.20
	acetonitrile	0.78
	butanol	12.03[a] (M^{-2})
	methanol	9.55[a] (M^{-2})
1,3-dimethylindole		
	butanol	44.57[a] (M^{-2})

[a] Longworth (10) was unable to reproduce these results and believes them to be due to the presence of water.

Reproduced with permission from ref. 9.

TABLE 7.14
Summary of the equilibrium constants for riboflavin species with various compounds[a]

Compound	pH	Concn (molar)	K_s	K_1	K_2	K_3
Tryptophan	7.06	.0089	.239	90	34 ± 3	126 ± 14
				42 ± 3	0	71 ± 11
Serotonin	8.86	.0051	.085	400	690 ± 18	405 ± 30
				0	95 ± 5	1.0 ± 8.0
Phenylalanine	7.10	.0050	.148	9 ± 5	0	22 ± 12
Tyrosine	7.10	.0019	.170	73 ± 12	0	130 ± 45
GSSG	6.75	.0050	.151	13 ± 4	0	26 ± 13
Pyridine	7.80	2.0	.0063	1.5 ± 0.1	2.0 ± 0.1	0
Nicotinic acid	6.90	.0053	.094	37 ± 14	21 ± 9	0
Picolinic acid	6.96	.0054	.120	10 ± 2	2 ± 2	0

[a] K_s is the equilibrium constant for the reaction of total oxidized riboflavin with the total reduced riboflavin to form the total amount of flavin semiquinone. K^1 is the equilibrium constant for the reaction of oxidized riboflavin with the organic additive to form a complex. K_2 is the equilibrium constant for the reaction of reduced riboflavin with organic additive to form a complex. K_3 is the equilibrium constant for the reaction of riboflavin semiquinone with the organic additive to form a complex. Reproduced with permission from Table 3, ref. 21.

TABLE 7.15

Summary of the equilibrium constants for flavin mononucleotide species
with various compounds[a]

Compound	pH	Concn (molar)	K_s	K_1	K_2	K_3
Tryptophan	6.96	.0049	.069	90	53 ± 5	25 ± 11
				56 ± 11	24 ± 14	0
Tryptophan	7.06	.0090	.099	90	69	80 ± 8
				13 ± 2	0	8 ± 7
Tryptophan	8.00	.0041	.102	90	71 ± 6	85 ± 16
				15 ± 5	0	10 ± 12
Serotonin	6.86	.0050	.069	400	600 ± 16	375 ± 45
	8.00	.0040	.127	400	600 ± 17	594 ± 34
Phenylalanine	7.10	.0050	.076	39 ± 14	21 ± 10	0
Tyrosine	7.10	.0019	.157	66 ± 12	0	170 ± 40
Imidazole	8.00	0.01	.108	1.6 ± 1.6	0	5 ± 2
Cystine	8.00	.00025	.121	0	16 ± 80	380 ± 290
GSSG	6.76	.0050	.141	7 ± 4	0	40 ± 11
	8.00	.0040	.151	0	6 ± 5	62 ± 15
Pyridine	7.76	2.0	.107	0	0018 ± 0.02	0.11 ± 0.02

[a] The equilibrium constants are comparable to those described in Table 7.14.
Reproduced with permission from Table 4, ref. 21.

TABLE 7.16

Association constants of D-amino acid oxidase complexes

Carboxylate	$K_c\,(M^{-1})$	ΔG° kcal/mole	ΔH°	ΔS° entropy units
valine	50			
leucine	1500			
isoleucine	71.5			
phenylalanine	150			
cysteine	370			
methionine	250			
acetate	30	− 1.5		
n-propionate	50	− 2.3	0 ∓ 1	7.8 ∓ 3.4
n-butyrate	475	− 3.7	0 ∓ 1	12.3 ∓ 3.4
n-valerate	4000			
n-caproate	2500			
n-heptanoate	680			
n-caprylate	500			
n-caprate	250			
i-valerate	53.5			
a-butenoate	43,000			
β-butenoate	1000			
cinnamate	20,000			
hydrocinnamate	320			

Determined in phosphate buffer at pH 8.3 and 25°C.
Reproduced with permission from Tables 1 and 2, ref. 26.

TABLE 7.17

Values of charge transfer band maximum of flavoprotein n-donor complexes

Carboxylate	$h\nu_{CT}$ (eV)
o-aminobenzoate	2.19
N-methyl-2-aminobenzoate	2.14
Δ^1-pyrroline-2-carboxylate	2.02
Δ^1-piperidine-2-carboxylate	1.97

In phosphate buffer at pH 8.3 and 25°C.
Reproduced with permission from Table 2, ref. 26.

TABLE 8.11

Properties of N-methylnicotinamide chloride complexes[a]

Donor	Association constant K (M^{-1})	Oscillator strength f	Dipole strength D (debye^2)	Obsd band maximum λ_{max} (nm)	Extinction[b] at λ_{max} $\varepsilon(\lambda_{max})$
Skatole	4.4 ± 0.2	0.053	3.9	320	1720
3-Indoleacetic acid	4.4 ± 0.2	0.037	2.7	314	1440
N-Acetyl-L-tryptophan	4.5 ± 0.2	0.035	2.5	314	1330
3-Indoleacetamide	3.4 ± 0.2	0.032	2.4	318	1320
N-Acetyl-L-tryptophanamide	4.0 ± 0.2	0.029	2.2	314	1180

$\Delta H^\circ = -3.5$ kcal/mole

Properties of N-methylisonicotinamide chloride complexes[a]

Skatole	3.8 ± 0.2	0.029	2.4	365	830
3-Indoleacetic acid	4.0 ± 0.2	0.020	1.6	355	670
N-Acetyl-L-tryptophan	3.9 ± 0.2	0.018	1.5	355	580

Properties of N-methylpicolinamide chloride complexes[a]

Skatole	1.5 ± 1	0.026		320	

Properties of the N-methylnicotinamide chloride–N-Acetyl-L-tryptophan complex in ethanol–water mixtures

EtOH (% v/v)	K (M^{-1})	$\varepsilon(350\ nm)$[b]
0	5.2 ± 0.3	830
1	4.5 ± 0.2	848
20	3.6 ± 0.2	800
40	2.2 ± 0.2	818

[a] 25°C, 1% ethanol.
[b] M^{-1} cm^{-1}.
Reproduced with permission from ref. 13.

TABLE 8.12

Association constants K and molar extinction coefficients ε of complexes between nicotinamide methochloride and various iodinated phenols[a]

Donor	$K(M^{-1})$	λ (nm)	ε
3-Monoiodotyrosine	$4^{c,\,d}$	400	190^b
3,5,3'-Triiodothyronine	4^d	400	285^d
3,5-Diiodo-p-cresol	6^c	460	218^c
3,5-Diiodotyrosine	4^b	410	$264^{c,\,d}$
Thyroxine	3.6^d	410	312^d

[a] These values are obtained by the Benesi-Hildebrand plot in excess of acceptor (0.1–0.5 M). The donor concentration is 2×10^{-3} M.
[b] Solution in Tris, pH 8.6, 0.05 M.
[c] Solution in Tris (pH 8.6, 0.05 M)—ethanol (15%).
[d] Sodium carbonate (pH 10)—dimethylformamide (10%).
Reproduced with permission from Table 1, ref. 34.

TABLE 8.13

Association constants K and molar extinction coefficients ε of complexes between diiodotyrosine and three electron acceptors[a]

Acceptor	N-Methyl-pyridinium chloride	N-Methyl-nicotinamide chloride	NAD$^+$
$K(M^{-1})$	0.4	4	40
ε_{410} mμ	250	264	55

[a] These values are obtained by the Benesi-Hildebrand plot in excess of acceptor (0.1–0.5 M). The diiodotyrosine concentration is 2×10^{-3} M (solution in Tris buffer, 0.05 M, pH 8.6).
Reproduced with permission from Table 2, ref. 34.

TABLE 9.5a

Equilibrium constants and spectrophotometric data for donor-acceptor complexes of egg lecithin with different acceptors in carbon tetrachloride at 24°

Acceptor	Conc. of acceptor $M \times 10^4$	Conc. range of egg lecithin $M \times 10^4$	K_c (M^{-1})	λ_{max}(nm)	$\varepsilon_{max} \times 10^{-4}$
Picric acid	0.65	3.69–18.45	1.25×10^4	360, 425	1.6, 0.91
m-Chloro CCP	0.47	3.55–17.75	3.50×10^3	362	1.24
Trinitrobenzene	8.03	5.82–17.46	2.44×10^2	—	—

Reproduced with permission from Table 1, ref. 35.

REFERENCES

1. Tegner, L. and Holmstrom, B. (1966). *Photochem. Photobiol.* **5**, 205.
2. Vaish, S. P. and Tollin, G. (1970). *Bioenergetics*, **1**, 181.
3. Slifkin, M. A. and Walmsley, R. H. (1970). *J. Phys. E.* **3**, 160.
4. Yagi, K. and Ohishi, N. (1970). *J. Biochem.* **67**, 599.
5. Okano, T. and Aita, K. (1967). *Yakugaku Zasshi*, **87**, 1243; *Chem. Abstr.* **66**, 88630.
6. Nakano, N. J. and Igarashi, S. J. (1970). *Biochemistry*, **9**, 577.
7. Montenay-Garestier, T. and Helene, C. (1970). *Biochemistry*, **9**, 2865.
8. Walker, M. S., Bednar, T. W. and Lumry, R. (1966). *J. chem. Phys.* **45**, 3455.
9. Walker, M. S., Bednar, T. W. and Lumry, R. (1967). *J. Chem. Phys.* **47**, 1020.
10. Longworth, J. W. (1968). *Photochem. Photobiol.* **7**, 587.
11. Stom, D. L. (1967). *Biofizika*, **12**, 153.
12. Shinitzky, M. and Fridkin, M. (1969). *Europ. J. Biochem.* **9**, 176.
13. Deranleau, D. A. and Schwyzer, R. (1970). *Biochemistry*, **9**, 126.
14. Bowd, A., Byrom, P., Hudson, J. B. and Turnbull, J. H. (1970). *Photochem. Photobiol.* **11**, 445.
15. Quinlan, K. P. and Fujimori, E. (1967). *J. phys. Chem.* **71**, 4154.
16. Quinlan, K. P. (1968). *J. phys. Chem.* **72**, 1797.
17. Quinlan, K. P. (1970). *Biochem. biophys. Acta*, **216**, 441.
18. Fory, W., MacKenzie, R. E., Wu, F. Y-H. and McCormick, D. B. (1970). *Biochemistry*, **9**, 515.
19. Karreman, G. (1961). *Bull. Math. Biophys.* **23**, 135; (1962) *Ann. N.Y. Acad. Sci.* **96**, 1029.
20. Knowles, A. and Roe, E. M. F. (1968). *Photochem. Photobiol.* **7**, 421.
21. Draper, R. D. and Ingraham, Ll. L. (1970). *Archs Biochem. Biophys. Acta*, **139**, 265.
22. Soma, M. and Yamagishi, A. (1970). *Biochem. biophys. Acta*, **205**, 183.
23. Land, E. J. and Swallow, A. J. (1969). *Biochemistry*, **8**, 2117.
24. Green, M. and Tollin, G. (1968). *Photochem. Photobiol.* **7**, 129.
25. Visser, J., Voetberg, H. and Veeger, C. (1970). "Pyridine Nucleotide-Dependent Dehydrogenases", p. 360 (Ed. H. Sund). Springer-Verlag, Berlin.
26. Yagi, K., Naoi, M., Nishikimi, M. and Kotaki, A. (1970). *J. Biochem.* **68**, 293.

27. Hemmerich, P., Nagelschneider, G. and Veegar, C. (1970). *FEBS Letters*, **8**, 69.
28. Bowd, A., Byrom, P., Hudson, J. B. and Turnbull, J. H. (1968). *Photochem. Photobiol.* **8**, 1.
29. Dudley, K. H., Ehrenberg, A., Hemmerich, P. and Muller, F. (1964). *Helv. Chim. Acta*, **47**, 1354.
30. Pullman, B. and Pullman, A. (1963). "Quantum Biochemistry." Interscience, London.
31. Douzou, P. (1968). "Molecular Associations in Biology" (Ed. B. Pullman). Academic Press, New York.
32. Cozzens, R. F. and Glover, T. A. (1970). *J. phys. Chem.* **74**, 3003.
33. Unzelman, R., Ludowieg, J. and Strait, L. A. (1964). *Experientia*, **15**, 507.
34. Mauchamp, J. and Shinitzky, M. (1969). *Biochemistry*, **8**, 1554.
35. Bhowmik, B. B. (1970). *Ind. J. Chem.* **8**, 299.
36. Saucin, M. and Van De Vorst, A. *Bull. Acad. Roy, Belg.* (in press).
37. Mantione, M-J. (1968). "Molecular Associations in Biology" (Ed. B. Pullman). Academic Press, New York.

Bibliography

The importance of charge transfer interactions in biology was first propounded by Szent-Györgyi in his book "Bioenergetics" and later in "Introduction to a Suvmolecular Biology," and "Bioelectronics".

The most thorough review of the topic is to be found in Volume 22 of "Comprehensive Biochemistry". A brief review is given by the Pullmans in "Quantum Theory of Atoms, Molecules, and the Solid State."

The Pullmans' book "Quantum Biochemistry" is a most valuable source book for the theoretical aspects of the subject.

Interesting and informative is the report of a symposium held in Paris in 1967 entitled "Molecular Associations in Biology" which contains articles by leading figures in this field.

Bullocks, F. J. (1967). Bioenergetics *In* "Comprehensive Biochemistry", Vol. 22 (Eds. Florkin, M. and Stotz, E. H.), Elsevier, Amsterdam.

Pullman, A. and Pullman, B. (1966). Charge Transfer Complexes in Biology, *In* "Quantum Theory of Atoms, Molecules, and the Solid State," (Ed. Löwdin, P.), Academic Press, New York.

Pullman, B. (1968). "Molecular Associations in Biology," Academic Press, New York.

Pullman, B. and Pullman, A. (1963). "Quantum Biochemistry," Interscience, New York.

Szent-Györgyi, A. (1957). "Bioenergetics", Academic Press, New York.

Szent-Györgyi, A. (1961). "Introduction to a Submolecular Biology", Academic Press, New York.

Szent-Györgyi, A. (1968). "Bioelectronics," Academic Press, New York.

Author Index

The numbers in parentheses are reference numbers and are included to assist in locating references when authors' names are not mentioned in the text.
Numbers in italics are those pages where references are listed at the end of chapters.

Compound Index

The main reference to a compound is indicated by italic numbers.

Subject Index

The main reference to a subject is indicated by italic numbers.

A

Acceptors,
  π-acceptors, 11
  σ-acceptors, 11
  shift in absorption band on complexing, 20, 195
Amino acid complexes, *52*
  n-donation from, 54, 100, 165, 170, 198, 227
Association constants,
  acceptor titration, by, 229
  Benesi-Hildebrand equation, by, 12, 116, 121, 151, 154, 156, 165, 173, 180, 214, 216, 217, 219, 223, 229, 230, 235, 245
  calorimetry, by, 42
  circular dichroism, by, 67
  competitive binding, by, 158
  dipole moment measurements, by, 43
  fluorescence quenching, by, 121, 142, 144, 158, 165, 200, 229, 230
  infrared spectra, by, 149
  kinetic methods, by, *45*, 111, 151, 156, 215
  nuclear magnetic resonance, by, 88, 124, 203, 216, 221
  partition methods, by, *39*, 87
  polarography, by, *39*
  potentiometric titration, by, 232
  solubility methods, by, *39*, 223, 228
  effect of solvent, 140, 229
  effect of substitution, 140
  effect of temperature, 176
  HOMO and, 144, 162
  inclusion of activity coefficient in, 88
  of solvated complexes, 13
  validity of, 165

B

Beer-Lambert Law, 7, 15, 50, 90, 110, 111, 173,

Benesi-Hildebrand equation, *11*, 12, 15, 17, 38, 53, 56, 61, 67, 116, 121, 151, 154, 156, 165, 173, 179, 180, 214, 216, 217, 219, 223, 229, 230, 235, 245
  for three interacting species, 14
  objections to, 14, 151
  Scott modification, 12
Brownian motion, 2, 234
Buffers, 47

C

Calorimetry, *42*
Carbonyl groups, 1, 10, 20, 74, 78, 79, 85, 91, 128, 134, 149, 163, 206, 233, 234
Carcinogens, 60, 76, 101
  as charge donors, 90
Cationic dyes, 89
Charge transfer bands,
  empiric characteristics of, *23*
  effect of solvent on, 24, 166, 173, 214
  effect of substituents on, 173
  extinction coefficient and, 7
  fine structure of, 24
  frequency of, 7
  multiple, 24, 100, 187, 229
  oxidative potential and, 186
  transition dipole of, 7
Charge transfer complexes
  classification of, *10*
  dipole moments of, 6
  effect of buffers, 47
  energy level diagram, 5
  excited state spectra of, 23, 91, 108, 137, 231, 232
  flash photolysis, of, 120
  fluorescence of, *21*, 78, 80, 81, 82, 88, 90, 107, 108, 109, 113, 136, 144, 154, 184, 207, 228, 229, 239, 240

267